Learning
in Science

THE AUTHORS

Roger Osborne) Co-directors of the Learning In Science Project based
Peter Freyberg) at the University of Waikato (1979-1984).

Beverley Bell is Education Officer for Secondary Science, Curriculum
Development Division, Department of Education.

Ross Tasker is Head of Science Education at Wellington Teachers
College.

Mark Cosgrove is Head of Mathematics and Science at Hamilton Teachers
College.

Brendan Schollum is Lecturer in Science at the Secondary Teachers
College, Auckland.

Learning in Science

The implications of children's science

Roger Osborne and Peter Freyberg

With
Beverley Bell   Ross Tasker
Mark Cosgrove   Brendan Schollum

Heinemann

Published by Heinemann Education, a Division of Octopus Publishing Group (NZ) Ltd, 39 Rawene Road, Birkenhead, Auckland. Associated companies, branches and representatives throughout the world.

© 1985 R. Osborne, P. Freyberg
First published 1985
Reprinted 1986, 1987, 1988, 1989, 1990(twice), 1991(twice) 1992

LIBRARY OF CONGRESS CATALOGING IN PUBLICATION DATA

Osborne, Roger.
   Learning in science.

   Bibliography: p.
   Includes index.
   1. Science — Study and teaching.   2. Science — Study and teaching
(elementary) I. Freyberg, P = S =
(Peter Stuart) II. Title.
Q181.083 1985   372.3'5   84-27915
ISBN 0-86863-275-9
SBN 435 57260 1

Printed in Hong Kong

Contents

Preface

This book is about children and adolescents, and about learning science. Young children and scientists have much in common. Both are interested in a wide variety of objects and events in the world around them. Both are interested in, and attempt to make sense of, how and why things behave as they do.

Einstein is reputed to have said that even physicists learn half their physics by the age of three. The young child repeatedly dropping a spoon from a high chair is beginning to learn about the physical world, albeit in a rather annoying manner for parents. Soon the child appreciates that if you let the spoon go it always moves to the floor — not upwards nor sideways; in fact there is an inevitability about many such happenings. He or she is beginning to learn about 'natural' phenomena and objects, and how to make sense of it all.

Language, too, usually develops early and even at pre-school age children have meanings for a large number of words. Included amongst these words are ones which have specific and agreed to meanings for the scientific community, terms such as 'animal', 'plant', 'light' and 'steam'. Sometimes these scientific meanings are subtly but significantly different from the meanings held by children.

Recent findings from a range of studies, including our own work on the Learning in Science Project in New Zealand, show that children bring to science lessons views of the world and meanings for words which have a significant impact on their learning. As a consequence children's ideas are influenced in unanticipated ways by science teaching.

In this book we explore these findings, analyse their significance for the teaching/learning process, and suggest both general and specific solutions to problems they identify for both learner and teacher. In Chapter 1 we introduce the reader to some children's views of the world and meanings for words which are unexpectedly different from those of adults in general and scientists in particular, and which influence children's subsequent learning in science. Because these views and meanings are often considered by children to be more sensible and more useful ideas than those presented to them by teachers, we refer to them as *children's science*. In Chapter 2 we look at some of the reasons why children's science is not influenced in the ways we might hope for in science teaching.

Part Two considers more specific problems related to children's science. Difficulties associated with language are considered in Chapter 3 where, as examples, we explore how children's views of 'animal', 'plant' and 'living' affect their understanding of lessons which hinge on the use of these terms. Then we discuss the need to change children's interpretations, and how this might be achieved. In Chapter 4 we consider how children construct their knowledge about the world from everyday events and occurrences; taking as our example children's views about force and motion. Here we also begin to explore the implications of what we have found out in our research for the designing of curricula and specific learning experiences. Chapter 5 considers further how children's ideas interact with those introduced by the teacher in science lessons and the consequent effects on learning. If science teaching is to influence how children think about the world as they have experienced it then what is learnt in school science must be about, or relatable to, that earlier experience. These issues are considered in the context of elementary learning in Chemistry. In the final chapter of Part Two we are concerned with activity-based lessons in science classrooms. We explore crisis points in such lessons, and this analysis leads to suggestions as to how the discrepancies between teachers' intentions for a lesson and learners' experiences can be reduced.

Part Three explores wider considerations relating·to the influence of children's science in the teaching and learning of science. Chapter 7 deals with some relevant aspects of learning and considers these in relation to the assumptions we as teachers frequently make about teaching and learning in general. In the light of our earlier discussion we then ponder as to what are appropriate aims for science education in the primary and secondary schools. In Chapter 8 we return to the classroom and to the implications for the classroom teacher of what we know about children's commonly-held ideas. New roles for the science teacher are explored and exemplified.

Are there important, even essential, components to a series of lessons which seek to modify children's strongly-held ideas? Part Four looks for answers to this question. In Chapter 9 various science teaching models which have been proposed by others are analysed. We introduce a teaching model which extends these ideas. In Chapter 10 the model is illustrated with a case study which outlines how children's views about electric current in simple direct-current circuits might be clarified, challenged, and modified.

Finally, in Part Five, we outline some of the implications of our research, as we see it, for the other areas of the curriculum and for teacher education. In Chapter 11 some broader implications across the curriculum are outlined, and in Chapter 12 we suggest how many of the ideas explored in earlier chapters can be presented and usefully discussed with teaching colleagues.

We have written this book mainly for practising science teachers and for use in pre-service and in-service courses in science education: nevertheless we believe that educators with other subject interests will find the ideas developed applicable to their own teaching. Our central concern has been with the teaching and learning of science at the 10-15 year old level, although children's science has implications for science teaching and learning at all levels of the education system. Those of us who are involved in the research which led to this book have found that our current teaching has been enriched by our experiences. *Learning in Science* is our attempt to share some of those experiences with a wider audience.

A note on research procedures

The following chapters record the results of a substantial body of research carried out under our direction or in collaboration with colleagues. Since the book is oriented principally towards teachers and curriculum developers we have not included full descriptions of research methodology, pupil samples, etcetera, in respect of each separate investigation. However, unless otherwise specified, all the studies referred to were conducted in New Zealand state primary, intermediate or secondary schools (ages 5 to 18 years). The pupils were drawn from intact broadbanded classrooms either randomly or as a teacher-selected representative sample. We have no reason to suspect that any of our pupil samples were biassed in terms of sex, ability or home circumstances.

Further details of research methodology and sampling will be found in the journal articles and working papers cited in connection with each investigation and, in particular, in Bell (1981b), Osborne and Gilbert (1980a, 1980b), Tasker (1980) and Tasker and Osborne (1983).

Roger Osborne
Peter Freyberg

March 1984

Science Education Research Unit
University of Waikato
Hamilton
New Zealand

Acknowledgements

Much of the research on which this book is based was funded by the New Zealand Department of Education. The Learning in Science Project (1979-1982 and 1983-85) was based on initial suggestions by Theo Ralfe and unfailingly supported by Les Ingham. Without their assistance the Project would not have materialised.

As directors of the Project we are indeed grateful to all those people who contributed so much of their time and energy to it. Ross Tasker was our first Project Officer, and along with Beverley Bell and Brendan Schollum led teams of researchers working on various aspects of the Project. Mark Cosgrove and Keith Stead always brought us back to consider deeper and wider issues. Local teachers have continued to support us in a whole variety of ways — by talking to us and by letting us talk to their pupils; by inviting us into their classrooms; and by working with us in the action-research phases of the Project. We can but admire the real professionalism of all those people who cooperated so readily with us, and particularly the members of our consultative committee who have given us sound and thoughtful advice at all stages of the research.

This book has been written by us in collaboration with those who did much of our field work during the first three years. They have added immeasurably to what we have learnt since then about the impact of children's ideas on the teaching and learning of science. Our special thanks go to them for making this book possible.

The authors and publishers also wish to thank reviewers in Britain, the United States and Australia for their comments on earlier drafts of this book. Particular thanks to Jack Lochhead and Howard Birnie in the United States for their assistance.

Part 1

The Problem

1 Children's Science

Roger Osborne and Peter Freyberg

Chemical equations. . . it took me ages to pick it up as I found it quite confusing. . . but having been taught by a teacher one way I tend to relate it in the same way but in my own thinking. . . in an exam I would probably get it wrong. You see when we are told to swot for a test we have got to swot in our book all the stuff the teacher's way. . . we go home and we try to learn that. . . but as soon as it hits our eyes it goes in our brain and it goes the other way. . . and so when we come to write it down and we think. . . and we write it down all our way. . . because of course it still means the same thing. . . there is no difference. . . but to the teacher there is a distinct difference between our way and the teacher's way. . . and the teacher's way is the right way. . . that's what I find hard.

15-year-old science pupil

Many children find it difficult to understand ideas put forward in science lessons, as this quotation clearly illustrates. But in our view the statement is more than just a plea for sympathy. The somewhat halting comments point us toward a major problem of learning science. Somehow the teacher's communication of his idea 'goes in our brain and it goes the other way'. It becomes the pupil's idea but somehow the pupil's idea is not the same as the teacher's idea. Many other children, too, have pointed up this problem to us. To quote another 15-year-old:

You know, teachers have got all that knowledge but we are thinking about it differently because there are so many ways you take something in.

During the past few years we have been involved in a research project which has focussed on the ways children are 'thinking about it differently' and on the impact that science teaching has on the ideas children bring with them to science lessons. Mostly, we have used individual interview procedures to explore pupils' meanings for words and their views of the world.

Many of the findings from this research will be discussed later in this book; in this introductory chapter we simply wish to illustrate the methods used, to exemplify the findings, and to state the major conclusions we have drawn from our own and similar studies.

Exploring children's ideas

Over recent years a wide range of procedures has been developed to investigate children's ideas about the world they live in. Many of these parallel the early studies of Piaget in that they involve individual interviews with a focus on natural phenomena such as movement or the boiling of water. The interviewer does not evaluate whether or not the child has an acceptable scientific concept but simply attempts to establish what are the child's ideas, however 'unscientific' these might be.

Two techniques have been used extensively in our research. The first, which we have called 'interview-about-instances' (Osborne and Gilbert, 1980a), is used primarily to explore the concept which a child associates with a particular label, for example, *plant*. Each child interviewed is shown a series of line drawings depicting various objects or episodes. The child is asked 'In your meaning of the word *plant* would you say that this is a *plant*?' or 'Would you say that there is a *plant* in this picture?' or, in the case of more abstract ideas 'What question would you have to ask me before you could decide if there is a *plant* in this picture?'. Once a positive or negative response has been obtained, the interviewer attempts to establish reasons for the response, for example. 'Why did you say that?' or 'Why is it not a *plant* to you?'.

For the investigation of the chosen words, diagrams are initially selected to illustrate a range of possible concepts which pupils might associate with each word. Then, following pilot interviews, further diagrams are added or deleted, to produce a set which provides a better reflection of the actual range of views observed.

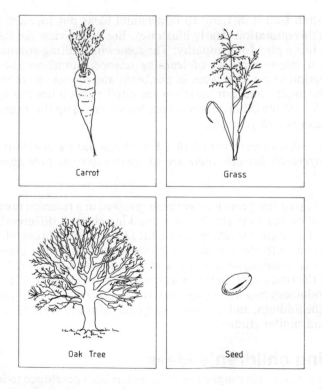

Figure 1.1: A sample of the cards used in an investigation of the word *plant* (after Bell, 1981a).

To illustrate the procedure, some of the diagrams Beverley Bell (1981a) used in her investigation are depicted in Figure 1.1. She established that, unlike biologists who classify living things into two main groups (plants and animals, with some exceptions), children often have a much more restricted meaning for the word *plant*. In a sample of 29 children, she found 10-year-olds, 13-year-olds and 15-year-olds who considered a tree was not a plant.

No, it was a plant when it was little, but when it grows up it wasn't, when it became a tree it wasn't.

10-year-old

Other children suggested that a plant was something which was cultivated, hence grass and dandelions were considered weeds and not plants by some 13 and 15-year-olds. Further, almost half the pupils interviewed considered that a carrot and a cabbage were not plants; they were vegetables. Over half those interviewed did not consider a seed to be plant material. Despite considerable exposure to science teaching many of the 15-year-olds held similarly restricted ideas to the 10-year-olds.

Our interview-about-instances procedures typically involve some 40 subjects to begin with. Each interview lasts about half an hour, is audio-taped, and as soon as possible after the event is transcribed by the interviewer. Subsequently the transcripts are analysed and the common elements and idiosyncrasies of children's ideas reported in working papers. Sometimes surveys are designed subsequently, to obtain an indication of the prevalence of commonly-found viewpoints at a range of age levels. These surveys are typically used with larger samples of pupils and, while not providing as conclusive results as would longitudinal studies (studies with one child repeated over a period of time), they give some indication of the impact, or lack of impact, of teaching and other experiences on pupils' views. Figure 1.2, for example, shows the percentage of pupils at various ages who consider an oak tree, a carrot, a seed and grass to be a *plant*.

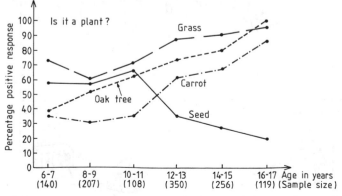

Figure 1.2: Results from a survey which investigated the prevalence of *plant* views at several age levels. See also the Note on Research Procedures, (page 2), for comments on pupil samples.

Using the interview-about-instances technique children's meanings for the words *force* (Osborne and Gilbert, 1980b; Gilbert, Watts and Osborne, 1982), *friction* (Stead and Osborne, 1981a), *gravity* (Stead and Osborne, 1981b), *animal* (Bell, 1981b), *living* (Stead, 1980), *energy* (Stead, 1981) and *electric current* (Osborne, 1981) have been investigated. As we will exemplify further when we discuss these studies in more detail, the ideas of many children do not appear to have been influenced in the ways we might expect by the teaching they have experienced.

The second technique used extensively in our research to explore children's ideas is a more flexible interview procedure which we have called interview-about-events (Osborne, 1980a). This method investigates children's views of everyday phenomena, such as situations involving the emission, reflection and absorption of light in and around the home, or involving the evaporation and condensation of water in the kitchen. The situations are presented to the children either through actual experience in the interview or by depicting them on cards.

For instance, we have investigated pupils' views about *light* using a set of cards, some of which are shown in Figure 1.3 (Stead and Osborne, 1980). Pupils were questioned about each picture. With reference to the candle (Figure 1.3a) they were asked *'Does the candle make light?', 'What happens to the light?',* and if appropriate in terms of earlier responses, *'How far does the light from the candle go?'* and *'How is it that the person is able to see the candle?'.*

Figure 1.3: A sample of the cards used in the *light* investigation.

The results of this study showed that children have a range of views about light phenomena, including some 9 and 10-year-olds with ideas which would be acceptable to the scientific community. On the other hand some older pupils, despite the formal science teaching they have been exposed to, still held ideas which suggested that this teaching had made no real impact on their fundamental views.

As evidence on this point consider again Figure 1.3a and the following pairs of responses to the question *'How far does light from a candle go?'*

First, two views which are likely to be quite acceptable to the scientific community even though they are rather vague:

'It goes about as far as it can go'.
'It goes out in rays and ricochets off objects'.

Another pair of answers — if one ignores the superficial differences of units — which would be considered less acceptable in terms of the scientific view of light:

'One metre at the most'.
'About one foot'.

Finally, a third pair:

'Just stays there and lights up'.
'Stays there'.

These pairs of responses were given by a 10-year-old and a 15-year-old, in each case — but not necessarily respectively! If it is difficult to determine which is which this simply emphasises the point we wish to make; despite the 15-year-olds having formally studied science for three years, including the topic *light*, it was common to find older students whose views were different from those of scientists but no different from those of many 10-year-olds. Yet these same 15-year-olds could when questioned define, reasonably well, terms such as *refraction* and *reflection*.

During our interviews, some children suggested that the distance the light travelled from the candle would depend on whether or not it was day-time or night-time. Invariably these pupils thought that the light would travel further at night. So we designed a survey to establish the prevalence of that view, two of the questions used being illustrated in Figure 1.4. Figure 1.5 shows the results, both for students who had studied the topic and for those who had not. While teaching may have had some influence on pupils' views about this phenomenon it can be seen that the effect is not great. Nor, we believe, is this just the result of poor teaching but rather of *what* is taught and *how* it is taught.

Returning to the original interview work, consider the following pupil's comment about the light from the electric heater (Figure 1.3e).

He could see it, and feel it, but the light does not reach him.

9-year-old

This view is not unique. Other children including an 11-year-old, a 13-year-old and two 15-year-olds from a sample of 36 9 to 15-year-old pupils also expressed it in varying words. Other interview cards elicited similar responses.

	The light from the candle:
	A: stays on the candle.
	B: comes out about halfway towards you.
	C: comes out as far as you but no further.
You are watching a candle burning during the day.	D: comes out until it hits something.

	The light from the candle:
	A: stays on the candle.
	B: comes out about halfway towards you.
	C: comes out as far as you but no further.
There is a power cut during the night. You are using a candle.	D: comes out until it hits something.

Figure 1.4: Two of a set of questions used to assess whether pupils considered that light travelled further at night than during the day.

Pupils were asked to choose the *best* alternative.

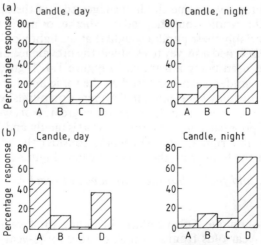

Figure 1.5: Survey results for the questions in Figure 1.4
   (a) 144 twelve-year-olds who had not studied light;
   (b) 235 thirteen-year-olds who had studied light.

How are we to explain the rather 'strange' ideas that some children have about light? From our study it became clear that children's ideas are strongly influenced by their egocentric or human-centred view of the world. Light from a candle, for example, is deemed to travel as far as any object which is obviously illuminated by it. If *they* (the children) can't see the illumination, then the light hasn't got as far as that. In the day time, objects more than about 0.5 metres from a candle do not appear illuminated by it, but the situation is different at night. A person in front of an electric heater does not appear illuminated by the light from the heater, although that person can 'see and feel' that the heater is on.

Most children would consider their ideas about light to be sensible, even commonsense, views. The study of light as a science topic often investigates its rectilinear nature and, for the instance, its relationship to concave mirrors, focal lengths and focal points; but some of the pupil's basic ideas about light, how far light travels, and how we actually see objects, may not even be mentioned let alone discussed. With children's ideas about these basic concerns being so different from the accepted scientific view perhaps it is hardly surprising that ex-pupils from our schools could be found who say:

Yes, we studied light. Mirrors and lenses and things. But I didn't understand what it was all about. I didn't have a clue.

adult interviewee

As a third and final introduction to our methods and findings we describe an interview-about-events investigation where pupils were given first hand experiences to discuss with the interviewer. The interview sequence involved a range of phenomena associated with the evaporation and condensation of water. At one point in the interview, each person was invited to observe water heating and then boiling in an electric jug (Osborne and Cosgrove, 1983). They were then invited to describe orally what was happening as the water came to the boil. When they mentioned the large bubbles in the water they were asked what the bubbles would be made of. Pupil's views included 'heat', 'air', 'air and water', 'oxygen and/or hydrogen' and 'steam or particles of water'.

The last answer, which corresponds most closely with the accepted scientific view, did not appear to be very commonly held in the sample of 8 to 17-year-olds we interviewed, so a survey question was designed to investigate the prevalence of each of the various ideas pupils had described to us (Figure 1.6a). The results are illustrated in Figure 1.6b. The pupils up to age 15 were studying general science. The 16 and 17-year-old pupils were specialising in one or both of Physics and Chemistry. Again there is a clear indication that what is taught or assumed does not match up well with what pupils learn or know!

Using the interview-about-events technique children's views about a range of other phenomena have been investigated, including the other *changes of state of water* (Osborne and Cosgrove, 1983), *dissolving* (Cosgrove, 1982), *burning and rusting* (Schollum, 1982), *soil* (Happs, 1982a), *rocks and minerals* (Happs, 1982b), and *weather* (Moyle, 1980).

a

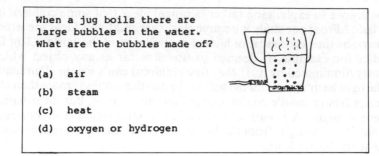

Figure 1.6: Investigating children's ideas about the large bubbles in boiling water: (a) the survey question and (b) the results obtained.
(16 and 17-year-old students were studying physical science.)

b

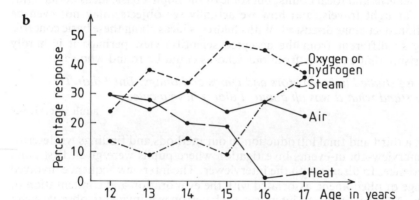

The nature of children's ideas

In this chapter we have illustrated our investigatory methods together with a few of our specific findings. Others will be discussed in later sections, together with the additional insights into children's ideas which we have gained through classroom observations. At this point, however, we draw attention to our general findings about children's ideas in science, which form a theme for the remainder of the book:

1   From a young age, and prior to any teaching and learning of formal science, children develop meanings for many words used in science teaching and views of the world which relate to ideas taught in science.

2   Children's ideas are usually strongly held, even if not well known to teachers, and are often significantly different to the views of scientists.

3   These ideas are sensible and coherent views from the children's point of view, and they often remain uninfluenced or can be influenced in unanticipated ways by science teaching.

We do not claim that these general statements are particularly original: they have been implied, if not stated in different ways, by several theorists from Piaget (1929) onwards. What we suggest, however, is that the consequences of such earlier findings have in fact been largely ignored by curriculum developers and teachers and, until recently, by other research workers in the field of science education as well.

Piaget's earlier work on animism, egocentrism, and the relationship between language and thinking, is still very relevant to science teaching and learning. In our view the broad questions of how children come to hold the ideas they do, and why such ideas are so difficult to modify, continue to be central research questions not fully explored within the narrow focus of much of the Piagetian-inspired research on stage theories and their implications.

This opinion is reinforced, for us, by the increasing amount of research on children's ideas in science which has been reported in the last few years in many parts of the world (for example, see Table 1.1). The import of most of this research is clear — unless we know what children think and why they think that way, we have little chance of making any impact with our teaching no matter how skilfully we proceed.

Finally, we have made no comment thus far about how children acquire their ideas prior to their formal teaching in science. It is sufficient here to observe that young children, like scientists, are curious about the world around them, and in how and why things behave as they do. Children naturally attempt to make sense of the world in which they live in terms of their experiences, their current knowledge and their use of language. Kelly (1969) suggests we are all scientists of a sort from a young age. The child-as-scientist develops ideas albeit tacitly, about how and why things behave as they do, which are sensible to that child. It is these ideas that we call *children's science* (Osborne, 1980; Gilbert, Osborne and Fensham, 1982). It is the similarities and differences between children's science and scientists' science that are of central importance in the teaching and learning of science.

Table 1.1: Examples of research which has explored children's ideas on natural phenomena (since 1976).

Specific Topics	
Kinematics:	Caramazza, McCloskey and Green, 1981; Trowbridge and McDermott, 1980, 1981; Jones, 1983.
Mechanics:	Leboutet-Barrell, 1976; Viennot, 1979; Champagne, Klopfer and Anderson, 1980; Hewson, 1981a; Gilbert, Watts and Osborne, 1982; Clement, 1982, 1983; di Sessa, 1982; Minstrell, 1982; White, 1983; Watts and Gilbert, 1983; McCloskey, 1983; Champagne, Gunstone and Klopfer, 1983.
Force:	McCloskey, Caramazza and Green, 1980; Osborne and Gilbert, 1980b; Sjoberg and Lie, 1981; Gunstone, Champagne and Klopfer, 1981; Watts and Zylberstajn, 1981; Watts, 1983a.

Energy:	Stead, 1981; Watts, 1983b; Duit, 1983; Solomon, 1983.
Friction:	Stead and Osborne, 1981a.
Floating and Sinking:	Rowell and Dawson, 1977; Rodrigues, 1980; Biddulph, 1983.
Gravity:	Gunstone and White, 1980, 1981; Stead and Osborne, 1981b; Watts, 1982.
Pressure:	Engel and Driver, 1981; Sere, 1982.
Heat:	Erickson, 1979, 1980; Tiberghien, 1980.
Temperature:	Stavy and Berkovitz, 1980; Strauss, 1981; Driver and Russell, 1982.
Light:	Guesne, 1978; Stead and Osborne, 1980; Eaton, Anderson and Smith, 1982; Goldberg and McDermott, 1983; Anderson and Karrgvist, 1983.
Change of State:	Andersson, 1980; Osborne and Cosgrove, 1983.
Electric Current:	Tiberghien and Delacote, 1976; Andersson and Karrqvist, 1979; Russell, 1980; Fredette and Lochhead, 1980; Osborne, 1981, 1983; Fredette and Clement, 1981; Hartel, 1982; Shipstone, 1982; Cohen, Eylon and Ganiel, 1983.
Chemical Change:	Schollum, 1982; Andersson and Renstrom, 1982.
Burning:	Schollum and Happs, 1982.
Particulate Nature of Matter	Novick and Nussbaum, 1978, 1981; Nussbaum and Novick, 1982; Osborne and Schollum, 1983; Brook, Briggs and Driver, 1984.
Earth Sciences:	Nussbaum and Novak, 1976; Nussbaum, 1979; Happs, 1982a, 1982b, 1982c; Klein, 1982; Sneider and Pulos, 1983.
Living:	Angus, 1981; Tamir, Gal-Choppin and Nussimovitz, 1981; Brumby, 1982.
Plant:	Bell, 1981a.
Animal:	Bell, 1981b; Bell and Barker, 1982.
Natural Selection:	Deadman and Kelly, 1978; Brumby, 1979; Kargbo, Hobbs and Erickson, 1980.
General Reviews	Driver and Erickson, 1983; Gilbert and Watts, 1983.

2 Science Teaching and Science Learning

Ross Tasker and Roger Osborne

Over the last decade curriculum developers in many countries have urged teachers to involve their pupils in more experimental work so that they can learn science by doing it. The teacher-directed or textbook-led experiment, with its associated group work, has thus become a common feature of many science classrooms. While there are problems associated with this work at a practical level — for example, lack of equipment, the need for technical assistance, and the organisational time required — we are here concerned with some of the more basic pedagogical issues involved in learning through 'hands-on' enquiry.

Our disquiet is perhaps best reflected in some typical secondary school teachers' comments. For example:

I have tried just about every approach (to classroom experiments) I can think of. . . and still there are so many in the class who I feel are not successfully working out what it is all about. They may try to think about it a bit, I don't know but they miss the point. . . I just think some of them are just not up to the stage where they can cope with our ideas. . . maybe some of them are not trying, it's hard to tell. . . you find that some of the most enthusiastic, eager, and willing types still miss the point.

science teacher

They always like the experiments but not always the ones you want them to be involved in. . . I mean, you know where you are going but it is hard to make them see where they *are going.*

young science teacher

Perhaps the most insightful comment was made to us by an experienced teacher who stated that '*they focus on things I would never dream of looking at, even.*'

As part of our Learning in Science Project we have investigated some of the problems and difficulties of learning science by watching and analysing the behaviour of pupils at work in science classrooms (see for example Tasker, 1980, 1981, 1982a). We observed them and recorded their class and informal small group verbal interactions on audio tape. We interviewed children about what they had done, about why they had done it and what they thought about it. We interviewed teachers about their intentions for a lesson, about their view of what actually happened, and about their perceptions of the outcomes of a lesson. Pupil activities and teacher actions were also described and analysed from the perspective of an interested but uninvolved observer. In addition we have drawn on our increasing knowledge of children's views of the world for insights into the reasons for specific actions and statements by pupils.

While we could illustrate our classroom-based work and findings from a whole range of content areas, we can best focus on what are, for us, some central problems by considering a single topic, simple electrical circuits. We have chosen this particular topic because we are able to support our classroom observational experiences with a considerable knowledge of children's ideas about simple electrical circuits based on extensive interview studies (Osborne and Gilbert, 1980a; Osborne, 1981, 1983).

A visit to two classrooms

Let us begin by describing what happened in two classrooms in which simple electrical circuits were being studied.

Situation 1:   In a class of 12-year-olds pairs of pupils are testing for electrical conductors
Each pair of children has been given a set of everyday items (screws, razor blades, a plastic comb, coins, paper, sellotape) and they have been asked first to list in their exercise books, on the basis of intelligent guesses, which of the items are conductors. Then they have to decide how they could find out if they are right, and finally to check whether or not their predictions are correct. The pupils have just finished making their predictions. . . .

> Teacher: *Now you've made your predictions. . . I want you to put up your hands please, and make a suggestion how you could test these things to see if they will conduct electricity. How could you do that?* (one boy holds up his hand) *Yes, Martin?*

> Martin : *Oh well, all you really need is a battery and a lamp and just. . .um. . .hook the wires up and have a break in the wires and put the thing down and put the wires into it.*

> Teacher: *OK, would you like to draw me a circuit diagram. . .Martin. Just to show me what you think we should do.*

Martin comes forward and draws a diagram on the board.

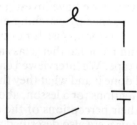

Figure 2.1: Martin's blackboard drawing.

What we consider Martin and the teacher were thinking is shown in Figure 2.2.

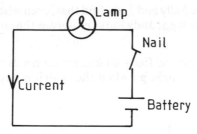

Figure 2.2: Martin's, and his teacher's thinking.

Teacher: *OK then, we'll try that out...now I want one person from your pair please to come up and collect the things.*

Sally and Judy, the pair of pupils in the class whom we focus on, construct their circuit (Figure 2.3). They are aware that they have a problem (the light will not come on) and when the teacher walks by they seek assistance.

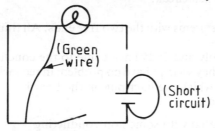

Figure 2.3: Sally and Judy's circuit.

The teacher immediately starts modifying their circuit to that expected.

Teacher: *Tut, tut, tut...you don't need that one.* (Removes short circuit.)

Judy   : *What's the matter?*

Teacher: *Well, that's because... look where you have got it connected up...you don't need that green wire to connect both of them...take the green wire out...Right, now we'll see if the circuit is working.* (Closes switch, and momentarily joins free ends so that the light flashes.) *OK, now you've got your circuit working.* (Figure 2.4)

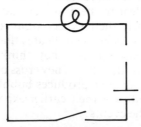

Figure 2.4: Teacher's circuit for Sally and Judy.

But after the teacher has gone Sally and Judy find that, even when the switch is turned on, the light still doesn't go! Judy looks closely at the circuit (Figure 2.4) and makes alterations.

Judy   :   *Oh, no wonder* (as she finds an extra wire needs to be added to make the light go when the switch is depressed). (Figure 2.5).

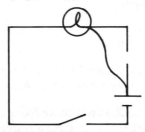

Figure 2.5: Sally and Judy's new circuit.

They then proceed to test items with the two free ends. All things tested are conductors!

Despite the fact that Sally and Judy have come to the conclusion that all the items are conductors, they seem to have no problem in reconciling this with the teacher's conclusions which are later put on the board.
The teacher's statements are:

• most metals unless coated with some non-conducting material will conduct electricity, for example gold, aluminium, copper.
• most plastic-based products, for example foam, sellotape, plastic bags, will not conduct electricity.
• products from nature, for example leaves, wood, fruit pips, will generally not conduct electricity.

Sally and Judy appear happy to write the teacher's statements into their books before they finish the lesson.

Situation 2:   Two 13-year-old pupils conducting an experiment labelled 'breaking up a compound'.

The two pupils manage to assemble the apparatus and note the bubbles rising from the anode. However it becomes clear to the observer that they are not really happy with the apparatus. Clearly their expectations are that both electrodes should give off gas. Their conversation indicates that, from their observations and in terms of their expectations, they think the battery connected to the non-gas-producing electrode is flat. They cross over terminals A and B, and note that the other electrode now produces bubbles, while the first electrode stops doing so. This reinforces their earlier expectation. The observer checks his interpretation of the events. . .

The following apparatus was the focus of the activity:

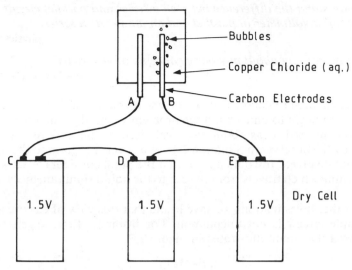

Figure 2.6: 'Breaking up a compound'.

Observer: *When you swapped the connections over... you did that because it wasn't working too well?*

John   : *Yes, it was working at one* (electrode) *but not the other. One of the batteries was too weak.*

Observer: *You had a weak battery?*

John   : *Yes, when we swapped the terminals over the good battery made the other terminal work.*

Observer: *If all the batteries were good would both of the terminals work at the same time?*

John   : *Yes.*

Observer: *Would there be an electric current in the liquid?*

John   : *Um... no... the electrodes are apart... there is no connection through there* (the liquid).

Observer: *No electric current in the liquid.*

John   : *Could be... but it hasn't got an outlet.*

The extent of the learning problems associated with the above two situations, particularly Situation 1, may be atypical but in our view the *nature* of the problems is not. However, electrical circuits are undoubtedly a good example of a science topic that pupils and teachers find difficult.

Electrical circuits... they get right through to the sixth form and they don't understand the difference between a series and parallel circuit... or why you put a voltmeter in parallel and an ammeter in series.

physics teacher

I just know nothing about any basic concepts of electricity.

teacher of 11-year-olds

Why do children find it difficult to understand electrical circuits? We would argue that to begin to understand the problem we need to understand some of the unexpressed ideas about electrical circuits that children hold. The probing by the observer in Situation 2 gives us some insight into the possible difficulties involved. It would appear that the pupil considers 'electricity' to be something which travels from the cell to the liquid simultaneously in both wires.

From other investigations we have found that this view of electric current in a simple circuit is not uncommon. For instance, Peter (aged 11) had constructed the circuit illustrated in Figure 2.7.

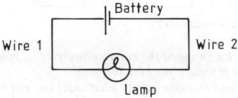

Figure 2.7: Peter's actual circuit.

The conversaton between Peter and the classroom observer included:

Observer: *How does electric current go?*

Peter   : *Well it goes from here* (the battery) *to here* (the bulb) *to make it go.* (Points to wire 1.)

Observer: *What does this wire do?* (Points to wire 2.)

Peter   : *That wire also helps because it puts it in there* (the bulb) *to help that one.* (Wire 1.)

When this was explored further with Peter, it was found that Figure 2.8 depicts the way he really thought that electric current flowed in the circuit.

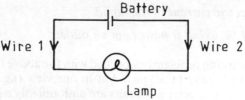

Figure 2.8: Peter's perception of his circuit.

Children's ideas about electric current

An understanding of current flow in a simple electric circuit would, again in our view, be a prerequisite to understanding other topics like series and parallel circuits, electric potential and electrical energy. We have interviewed over 100 children in England (Osborne and Gilbert, 1980a), New Zealand (Osborne, 1981) and California (Osborne, 1983) about their ideas on electric current in a whole range of situations.

As part of this interview-about-instances schedule children were asked, 'Where does the electric current which goes into a lamp go to?'

Here are three examples of pupils displaying sound scientific ideas:

It goes back to the power source.
12-year-old science pupil

It carries on its circuit.
14-year-old science pupil

It goes on around the circuit.
16-year-old physics pupil

However, here are some other answers to this question:

It is transformed as energy. . . heat and light energy.
17-year-old physics pupil

It is used up.
12-year-old science pupil

It is used up to create heat and light energy.
15-year-old science pupil

It goes out as energy in the light.
16-year-old physics pupil

It turns into light, probably used up.
14-year-old science pupil

It gets burnt up.
11-year-old science pupil

It would appear from these statements that pupils do not always have the expected scientific view of current flow — but they do have ideas. Furthermore, the ideas are often similar, despite considerable differences in age and learning experiences. The kind of conceptions which children hold prior to formal teaching is also indicated by our observations of 40 children, aged between 8 and 12 years, who in an individual interview situation were given two wires, a battery and a bulb and asked to make the bulb light up (Osborne, 1983).

Of these children 22 tried to get the bulb to light up using one of the methods shown in Figure 2.9 (a-d).

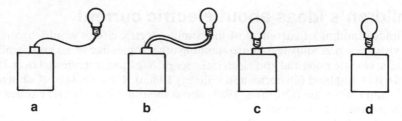

Figure 2.9: Attempts to get the bulb to light (Type I).

A further 11 children tried using a method shown in Figure 2.10 (a or b).

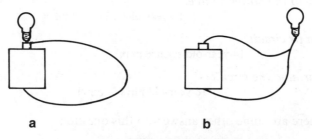

Figure 2.10: Attempts to get the bulb to light (Type II).

Finally, six children actually got the bulb to light by using one of the methods in Figure 2.11 (a or b). Only one child refused to attempt the problem.

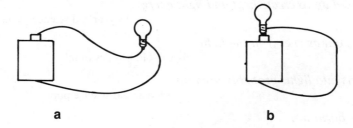

Figure 2.11: Attempts to get the bulb to light (Type III).

How *do* children think of electric current in a simple circuit? When one considers the way many young children try to wire up a battery and a bulb it is obvious that many of them have what we call a Model A view of electric current. That is, that electric current flows from the upper terminal of a battery to the base of the bulb. One wire (or contact) is all that is needed — any other wire is either unnecessary or just an extra.

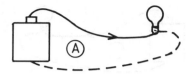

Figure 2.12: Model A, 'There is *no* electric current in the wire attached to the base of the battery'.

From our interview work we find that some children retain this model even when they have seen that it is necessary to have two wires connected to two different parts of both battery and bulb before the bulb will glow. Pupils in this category have told us that the other wire is a 'safety wire', or it's 'just required to get the bulb alight', rather like a catalyst.

Returning to the interviews with the 8 to 12-year-olds we referred to earlier: once these children were shown that it was necessary to have *two* different sets of contacts before the bulb would light 35 out of the 40 children described what was happening with what we call a Model B view of electric current (Figure 2.13).

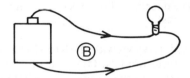

Figure 2.13: Model B, 'The electric current is flowing in a direction *toward* the bulb in *both* wires'.

In this view electric current is seen as travelling in *both* wires *from* the battery *to* the bulb. The model itself is not really such a silly idea. In fact it appears to be one held around the beginning of the 19th Century by a certain French physicist called Andre Marie Ampere, 1775-1836.

(Two electricities are carried) in such a way that there results a double current, one of positive electricity, the other of negative electricity, starting out in opposite senses from the point where the electromotive action arises, and going out to re-unite in the parts of the circuit remote from these points. . . it is this state of electricity in a series of electromotive and conducting bodies which I name for brevity electric current.

quoted in Hurd and Kipling, 1956, p.221

The remaining five children out of the 40 interviewed (Osborne, 1983) described what we call a Model C view of current — that is, the direction of the electric current will be as shown in Figure 2.14. The current circulates but there is considered to be less current in the 'return' wire.

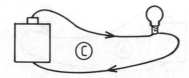

Figure 2.14: Model C, 'The direction of the electric current is one way. The current will be *less* in the "return" wire.'

In addition to these three views, however, we have found through other interviews that some children hold the scientists' view, even from quite a young age, with the direction of the electric current as shown in Figure 2.15, the current being the *same* magnitude in both wires.

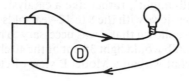

Figure 2.15: Model D, 'The direction of the electric current is one way. The current will be the *same* in both wires'.

The final section of our interviews with the 8 to 12-year-olds provided further confirmation of their various models. Once the children had discussed their ideas about the electric current in the circuit, they were presented with the four different models discussed above, and were told that these were various ways in which people had talked to us about the current flow in the battery-bulb circuit. They were asked to comment on these models.

Typically, when shown Model A, many children said *'That is the way I used to think about it, until I realised that the second wire was needed'*. Some children considered this was *still* a sensible way to think about it, even though they realised that the two wires were required. As one 10-year-old said, *'Yes, the bulb needs all the electricity.'*

When shown Model B many children said, *'Yes that's the way I think about it now'*. As one 11-year-old put it: *'The currents clash in the bulb.'* When shown Model C a number of students thought it was a reasonably sensible idea. One 11-year-old girl said, *'Yes, it makes more sense to me why the return path is needed. I don't think electric current all gets used up.'* (When shown Model C, one first-year University physics student recently commented *'Yes... there must be some loss due to the fact that the light is glowing causing a current loss.'*)

It was a different picture with Model D (the physicist's model), however. Almost all the children rejected it as being either completely unintelligible or implausible. For example:

(If it was Model D) then how could it keep lighting (the bulb)?

8 year-old-boy, 9 year-old-girl

It is not right. . . some electricity is used up in the bulb.

9 year-old-girl, 11 year-old-boy

It is not D. . . if it was the battery would never get worn out. . . but that is impossible because it does.

12 year-old-boy

(Even a first year University physics student, shown Model D, commented, *'No, that's not right, you don't get something for nothing. . . there must be a drop in current.'*)

To estimate the prevalence of these various views of electric current, the question shown in Figure 2.16 was asked of a wide sample of students from New Zealand schools. The results of this survey are shown in Figure 2.17, and indicate how children, even at age 11, have a range of differing views. The results also illustrate clearly that the idea of current in simple circuits having the same 'strength' in all parts of the circuit, has not been learnt effectively even by older children. The increasing prevalence of the Model C view, up to the age of 15, is indicative of children interpreting new information in the wrong way. Perhaps it is no wonder that 15-year-old science pupils have difficulty with Ohm's Law, if half of them have a Model C view of current flow in a simple circuit. The results at age 16, from physics classes, and from 17 and 18-year-olds (who were final year secondary school physics and first year university physics students) may, of course, indicate simply the self-selection into physics classes of students who hold the Model D view, rather than effective teaching!

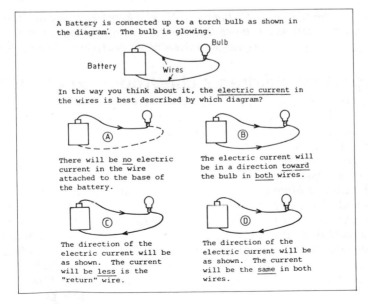

Figure 2.16: Survey question about electric current.

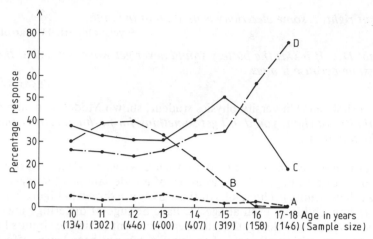

Figure 2.17: The results of the survey of electric current 'models'.

In some discussions with some of these older children we see how they reinforce their existing view with newly taught knowledge. For example:

The positive current goes one way and the negative current goes the other way.

15-year-old science student choosing Model B

Resistance slows it down so there is less current in the return path.

17-year-old physics student choosing Model C

I chose C because if V = IR then I = V/R. Therefore if voltage remains the same and resistance is more then the current after must be less.

18-year-old university physics student choosing Model C

Interpreting children's classroom experiences

Having considered some of the ideas about electric current that children typically hold, we are now in a better position to analyse what was happening in the two classroom situations we described in the first section of this chapter. Both illustrate how the possibilities of learning what the teacher intends can be severely restricted by the conceptions which pupils inevitably bring with them to the classroom.

In the case of *Situation 1,* it is clear that Sally and Judy did not really understand what an electrical circuit involves — a continuous channel for the flow of current. Hence the idea of using a gap in the circuit to test whether or not a particular substance is a conductor was beyond their comprehension. The only ideas they possessed to build on appeared to be their knowledge that batteries, wires and bulbs, joined together, somehow cause the bulb to light up: under these circumstances it is not surprising that they proceeded in a mainly haphazard manner until they achieved *their* main goal, getting the bulb to glow.

As a consequence we can observe a major difference between the teacher's intentions and the learning which was actually taking place. This difference was, in turn, the direct result of a number of disparities:

- There was a disparity between the *ideas* children brought to the lesson and the ideas the teacher assumed that they would bring to it.
- There was a disparity between the *scientific problem* the teacher would have liked the children to investigate and what they took to be the problem.
- There was a disparity between the *activity* proposed by the teacher and the activity undertaken by the children, despite considerable teacher intervention.
- There was a disparity between the children's *conclusions*, and the conclusions proposed by the teacher.

What happened in Situation 1 can be considered from another perspective. Wittrock (1974) believes that pupils as individuals will inevitably construct their *own* purpose for a lesson, form their *own* intentions regarding the activities they will undertake, draw their *own* conclusions and carry these through in their subsequent thinking. Whether or not these constructions are similar to, or quite different from, those intended by the teacher will depend on a range of factors. What has to be accepted is that teacher's intentions cannot be transferred directly into pupil intentions. Teachers must contrive learning situations in such a way that the mental constructions made by pupils — what the lesson is about, what is to be done, and what can and is to be learnt from it — correspond with their own intentions. In our view it is by appreciating, amongst other things, the perceptions that the learner is bringing to the lesson that teachers can reduce the disparities between teacher intentions and pupil's learning which were so evident in Situation 1.

Situation 2 provides another clear illustration of how children bring ideas with them to the learning situation which dramatically influence their learning. If, as seems likely, these pupils held a Model B view of electric current in a simple circuit, then it is not surprising that their interpretation of what they observed led them to conclude that one of the batteries was flat. Moveover their Model B view would have been *reinforced* by their interpretation of what happened when the wires to terminals A and B were reversed; now the 'flat' battery did not provide the first electrode with enough electricity to produce the gas.

These two illustrations are typical both of the type of investigation in science classrooms that we have been engaged in and of the discrepancies in teacher/pupil intentions that we have been observing. Such findings are supported by the work of others (for example, see Driver 1981, 1983). The discrepancies can be critical in terms of the impact which a learning experience has on children's viewpoints. In Chapter 6 we will analyse more systematically and in greater detail these contrasts between the learner's point of view and that of the teacher. We will elaborate further on the reasons why children's ideas often remain uninfluenced or are influenced in unanticipated ways by science teaching, and indicate ways in which our teaching intentions might be better reflected in our pupils' learning.

Part 2

Towards Specific Solutions

3 Language in the Science Classroom

Beverley Bell and Peter Freyberg

Teacher: *Is a person an animal?*

Class   : *No!* (chorus)

Jane   : *Yes it is!*

Class   : *No it's not!*

Jane   : *Yes it is. We are all animals.*

Teacher: *Why do you say that, Jane?*

Jane   : *Well, we are all kiwis,*[1] *aren't we?*

from a New Zealand class of 8-year-olds

The teacher of this class was discussing whether people are animals because he had been alerted to the fact that many young children do not consider people to be animals as a biologist would. Both Jane and the teacher accept the biologists' view that a person is an animal. Although they both agree that 'a person is an animal', they do not, however, necessarily have the same or even similar conceptions underlying that statement. It is this problem on which we wish to focus in this chapter.

Children's meanings for animal and living

The word *animal* is used extensively in science textbooks and science lessons. If children are thinking of different categories of living things from that intended by the teacher, then there is room for a wide range of conceptions, as we have discovered.

Using the interview-about-instances approach discussed earlier we have undertaken several investigations of pupils' meanings of the word *animal*, for example (Bell, 1981b; Bell and Barker, 1982). The exemplars and non-exemplars pictured on cards and shown to pupils were, in order; seagull, cow, spider, worm, grass, cat, mushroom, rock cod, boy, frog, snail, elephant, snake, fire, lion, whale, car, tree, and butterfly. Examples of the pictures are shown in Figure 3.1 also.

To a biologist only the grass, mushroom, fire, car and tree would not be animals. However only four of the 39 average ability pupils, ranging in age

---

1 New Zealanders are sometimes known as kiwis, after a flightless bird unique to this country.

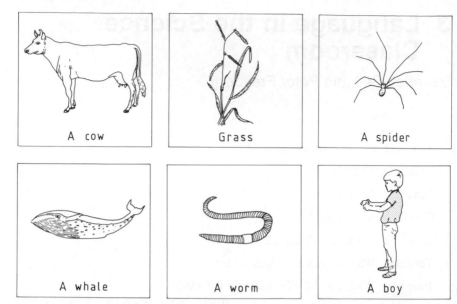

Figure 3.1: Some of the interview cards used to investigate children's meanings for the word *animal*.

from 9 to 15 years, categorised the exemplars in this way. Many of the pupils considered only the larger land animals, such as those found on a farm, in a zoo, or in the home as pets, as animals. Reasons for categorising something as an animal, or not doing so, included the number of legs (animals are expected to have four), size (animals are bigger than insects), habitat (animals are found on land), coating (animals have fur), and noise production (animals make a noise).

IS IT AN ANIMAL?				
	11 year olds (N = 49)	Primary teacher trainees (N = 34)	Experienced primary teachers (N = 53)	University biology students (N = 67)
Cow	98%	100%	100%	100%
Boy	57%	94%	96%	100%
Worm	37%	77%	86%	99%
Spider	22%	65%	86%	97%
Grass	0%	0%	0%	0%

Table 3.1: Percentage responses of various samples to 'is it an animal?'

Surveys were devised to establish the prevalence of these views amongst more representative samples. Table 3.1 shows the proportion of samples of primary teachers in training, experienced primary teachers, and first year university biology students, who considered a cow, boy, worm, spider, and grass to be an animal. All these people were asked to respond while they were engaged in a biology or science course activity. In Figure 3.2 the proportion of responses for representative samples of 5 to 15-year-old pupils as well as 16 and 17-year-old biology students is provided. The U-shaped curves for whale, worm and spider have been found with other samples and are worthy of some discussion.

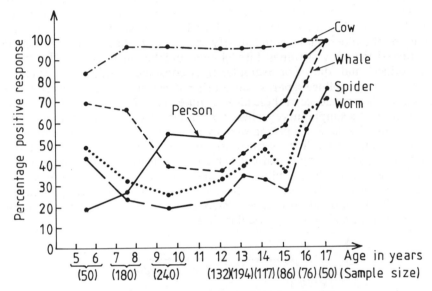

Figure 3.2: Representative sample response to 'is it an animal?'

Interviews we have carried out with 5-year-olds indicate that children of that age have a relatively simple classification system. Living things consist of plants and animals. Older children, however, learn about insects, mammals that live in the sea, arachnids, and so on. For some of these pupils at least their concept of animal becomes restricted to the land mammals. One pupil we talked to said:

...before we started on science, when we were in the younger school, when they said animals, you sort of thought (of) all animals. You didn't sort of class insects as insects. Everything was just an animal. It wasn't human. But now...they take things away. That group is insects...and mammals or whatever...When you get older you are told that they are insects, and they are mammals...and at this age you are told that everything is just an animal or a plant...they're wasting time where they say that they are splitting everything up. And then they are just putting everything back together at the end again.

13-year-old

It would appear that this pupil, like others, had not comprehended that insects and mammals can be considered as subsets of the group animal, and not as parallel groups.

Moreover, it is the narrower meaning of the word 'animal' which is often used in everyday language, even by scientists. For example, when we read the shop sign 'No animals allowed', we interpret it to mean that no pets are allowed inside. Although scientists call people animals, people are not animals in the everyday sense of the word and so we can enter the shop to buy food. Similarly animals go to the vet, while humans go to doctors!

In another study (Stead, 1980) we investigated pupils' meanings of the word *living*, again using an interview-about-instances approach. A surprising number of pupils considered fire, clouds and the sun to be living; furthermore, some of the older pupils could justify their categorisation in terms of the characteristics of living things. Fire, clouds, and the sun all 'move', 'breathe', 'reproduce' and 'die'. For example, fires consume wood, move (flicker), require air, reproduce (sparks cause other fires), and give out wastes (the smoke). The prevalence of these ideas is indicated in Figure 3.3, which shows the results of a simple survey where respondents were asked whether or not a fire, a moving car and a person were living. The metaphoric, rather than the scientific, meaning of living is the most common usage in everyday language. We say the fire is living because it behaves *as if* it were living. We talk of a 'live' wire and the 'living' bible. Like the word *animal*, the word *living* has two common meanings — a scientific one and one used in everyday language. The problem for pupils in science classrooms is to learn to tell which of these meanings is intended on a particular occasion.

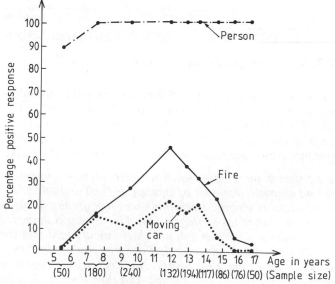

Figure 3.3: Representative sample response to 'is it living?' (16 and 17-year-old pupils were studying biology).

Constructing the intended meaning

When a teacher talks to her class, draws a diagram on the blackboard, discusses a chart on the wall, or asks pupils to read a textbook, her intended meaning — or that of the textbook author — is not automatically transferred to the mind of the pupil. Each individual in the classroom constructs his or her own meaning from the variety of stimuli, including the specific words read or heard, which are present in the learning environment.

How similar the constructed meaning is to that intended by the teacher — indeed if any meaning is constructed at all — depends on the way a pupil copes with the language we as teachers use so freely as our main means of instruction. Here are some examples of the differing ways pupils and teachers can react to the language of the science classroom.

Ignoring teacher talk: If the teacher's language includes words unfamiliar to pupils, which are not explained in the pupil's language, comprehension of what is being said will not occur. Pupils simply fail to construct meanings from the teacher's flow of words:

It would have been good if we had known what the words meant.

16-year-old

An associated problem highlighted by Edwards and Marland (1982) is that pupils frequently start to construct a meaning from what is being said but the process of construction triggers associated ideas from the memory store which have no bearing on the line of argument being presented by the teacher. The pupil drifts off to another world. At the secondary school level, for example, a science teacher may be talking about the world of theory while his pupils drift back to the world of reality — to past experiences, images, and episodes.

Of course this inability to construct appropriate meanings in the classroom does not apply only to teacher talk:

They (the teachers) *just wrote it on the board and you wrote if down in your little book. . . or you copied it out of your science textbook, which meant nothing.*

ex-pupil

Where the teacher's own concepts are inadequate, there is unfortunately a greater likelihood that he or she will consciously or sub-consciously try to obscure his or her lack of understanding by the use of technical language, whether it be verbalised, written on the board, or referenced from the textbook.

Noises which sound scientific: Where the teacher demands that pupils use the teacher's language it is possible for pupils to play the game. As Barnes (1976) has pointed out, an insistence by teachers on the correct use of words is considered by many children as an instruction about words rather than about concepts. Our own classroom observations have often confirmed this. One pupil we observed, for example, was told *'Well done, you are thinking like* *a scientist now'*, by a teacher whose own concepts of the material concerned were at the best fuzzy. This pupil had included some appropriate scientific terms in what we considered to be an otherwise incoherent response. Children can learn to make noises which sound scientific and to spell and pronounce otherwise meaningless words.

We actually had to learn how to spell photosynthesis which took ages for the whole class to spell it right. . . but even the teacher couldn't say what photosynthesis meant.

16-year-old talking about science at age 12

Ignoring pupil-talk: Irrespective of the teachers' scientific understandings, their use of unfamiliar language allows them to be in control of the situation, provided always that they can maintain a certain level of pupil interest in what is going on. Pupils who attempt to interact with the teacher using their more hesitant and less eloquent expressions are automatically at a disadvantage. Such attempts can be easily devalued, and even ignored by teachers. A frequent change of topic also makes it hard for pupils to interact.

You just start to get to know what you're talking about and they (the teachers) *change it* (the topic). . . *you forget everything that you know. . . in the end you do not know what you are doing.*

ex-pupil

An English researcher described a lesson he observed which could apply to a number of our own observations of science lessons;

Far from helping them to bridge the gulf between his frame of reference and theirs, the teacher's language acts as a barrier. . . they are left with their own first hand experience. . . the state of the other less articulate members of the class can only be guessed at. . . the teacher, frightened by his sudden glimpse of the gulf between them, hastily continues with the lesson he has planned. . . the teacher teaches within his frame of reference, the pupils learn in theirs. . . taking in his words which mean something different to them, and struggling to incorporate this meaning into their own frame of reference.

Barnes, 1969, p. 28-29

The unidentified mismatch: When the language of the teacher involves familiar words used with specialist meanings in the science classroom, particular difficulties can occur because both pupil and teacher may be unaware of, or more importantly unable to identify, the source of the problem. Words such as *animal, plant, living* and even *consumer*, each have two or more subtly different meanings, some of which are compatible with other sections of the message but which are not the meaning intended by the teacher or textbook author. For example the statement, 'animals are living things' could be said by any teacher, agreed with by all pupils, but could result in pupils constructing a widely different range of meanings depending on whether they are using a scientific meaning of animal and living, or a meaning used in everyday language. The word *consumer*, which has a different meaning in science to that which it has in economic studies, provides another example. In biology, the word consumer means a living thing that eats ('consumes') another living thing as food. In economic studies the word means a 'user' of a product. Problems can occur if the economic studies meaning is unconsciously taken up in a science lesson. For instance one pupil who was discussing the sentence 'all animals are consumers', expressed the following point of view:

All animals eat things, and if it doesn't eat anything it's not an animal, like a flower doesn't eat things so it's not a consumer. Well, it is a consumer but not of food. Doesn't eat animals or plants. Gets the minerals in the soil and goes up. (So would you call plants consumers?) Yeah, sort of. Yes, because they use the rays of the sun and minerals in the soil.

13-year-old

A biologist would not call plants consumers as they make their own food by photosynthesis.

The identified mismatch: Sometimes, pupils are aware of a mismatch between the meaning for a word used by the teacher and that which they use themselves, but they continue using their own meaning despite the acknowledged difficulties. For example, one pupil we talked to about the meaning of the word animal, said:

(Which of the two meanings do you think you use the most?) The everyday meaning. (Which one would you use with Mr P———— in the science lesson?) The everyday meaning — it's easier to remember. (Which one do you think Mr P———— uses?) The scientific one. (Do you think there are any problems in that?) Yep, because what he's saying we don't really understand.'

13-year-old

Furthermore, recognising an author's or teacher's intended meaning does not always involve accepting and using it. For instance:

...I still think spiders are insects, not animals. Don't know why. I just don't feel a spider's an animal...I don't really agree with that one either. — I don't agree with it because I don't think an earthworm's an animal either.

13-year-old

Having apparently constructed the author's intended meaning, another pupil used it to answer a survey test in the science lesson. However, when interviewed later the pupil stated:

(A person) is an animal as it eats, both meat and vegetables. A person does that. (Does that make sense to you — a person is an animal?) Not

*really. (A scientist would call a person an animal. You've ticked 'Yes'
there. Do you think you'd think like a scientist now and call them an
animal?) Not really — you'd get your head thrashed in for one. You just
don't think of people as animals.*

13-year-old

This pupil played the 'game' of getting the correct answer but was unable
or unwilling to reconcile the construction with his existing knowledge. To call
another human being 'an animal' had strong negative connotations!

'Ordinary' words.   It is not just the technical and scientific words that can
lead to communication difficulties. Common words can also have a range of
meanings. Take for example the word *make* in the sentence 'Plants can make
their own food using the sun's energy but animals are unable to make their
own food'. A 13-year-old pupil made this comment on the sentence:

*Well, animals don't have hands and things that they can make their own
food with, because they've just got four legs. That's very awkward. They
aren't able to do things like we are because we have fingers and hands to
help us. (So people can make their own food?) Yeah...they could grow
their own crops, they could buy things or like grow things to make
things... biscuits and that. And animals just have to use their mouths.
Unless it's a monkey.*

Not only is she using *make* in a way never intended by the author, but she
is also using *animal* in the everyday, not scientific, sense. Her understanding
of the sentence must have been very different from that which the author
intended in this instance.

Solving the language problem

We have identified several problems involving language meanings in science
classrooms, not only because the words used can be unfamiliar to pupils but
also because even simple words can have different meanings in differing
contexts.

To help pupils appreciate these differences in meaning we have frequently
placed pupils in small groups and asked them to classify interview-about-
instances cards as a group. The interview-about-instances cards dealing with
animals, when used with 14-year-olds for example, can result in an interesting
and lively debate, but will not necessarily lead to the biologist's viewpoint being
finally accepted. The following extract is taken from the discussion by four
14-year-olds who could not agree about whether or not a spider was an animal.

Rangi : *It's an insect... it is not* (an animal).

George: *But it contains blood and all that...so... it's got eyes and
it's got a mouth.*

Jane   : *Six legs...*

Maria : *It is an insect and an insect is not an animal.*

(George is most unhappy about putting the spider in the non-animal pile.)

Jane : *Do you agree* (it is not an animal) *or not?*

George: *No, I do not agree. . . you persuade me it is not an animal.*

Maria : *Well, it is an insect, isn't it?*

Jane : *You know in science and that they* (the teachers) *always class spiders as insects.*

George: *No they don't. . . they have another name for them.*

Jane : *I have never heard them* (the teachers) *say they are animals.*

Maria : *It seems too small to be an animal. . . You think of an animal to be large.*

George: *I say you can get ultra-small animals.*

Jane : *Yes. . . 'cos guinea pigs aren't all that big.*

George: *You can get big spiders.*

Maria : *Not quite as big as guinea pigs.*

George then decides to accept the views of the other members of the group, at least for the moment, and the group were then able to move on to the next card. But who really knows what George's conception of an animal was by that time — or that of the rest of the group?

Exercises such as these — with similar possible outcomes — may seem rather hazardous teaching devices. In our view, however, they have considerable value *provided* they are followed up by the teacher. Here we have children verbalising their ideas in science. All too often the discussion between pupils in science classrooms is confined to adminstrative details, such as who will go and get the equipment, where graph paper is, who will measure the time, and so on. All too rarely do we find pupils talking about their own conceptualisations of the scientific ideas underlying the 'scientific' activity they are meant to be engaged upon. Also, in our opinion, if learners are to change their views, or to appreciate what is being taught, it is important that they are clear about their *own* views. If they remember little else it is possible that the four pupils just described may appreciate that there is a range of views about animals and will consequently be more interested in the scientific viewpoint, and will subsequently be prepared to consider how that viewpoint differs from their own.

In our own teaching about animals, having identified the necessity for explicating the biologist's concepts of animals even with 16-year-old biology students, we have concentrated on the proposition that all living things may be classified into two main groups — plants and animals. If a living thing is not a plant, then it must be an animal, and vice-versa. We have explained how birds, insects, mammals are subgroups of animals. We have done this using tree diagrams, venn diagrams, and verbal descriptions. We have also explained the differences between the everyday usage of the word and the biologist's usage. We have exemplified the points by involving pupils in a variety of tasks, for example crosswords, card games, dice games, which make use of these ideas (Bell, 1981c). Similar tasks have been used to teach about plant and living. Only when the biologist's meaning is accepted as well as recognised, we feel, will much of the related biological teaching to which pupils are exposed make sense to them and more subtle misinterpretations be avoided later.

Language and learning

Some of the problems which occur when pupils employ a different meaning of a word to that of the teacher have now been outlined. But does this potential mismatch really influence learning to any significant degree? To answer this, we can cite as an example a study by Bell and Barker (1982) in which 13-year-olds were introduced to the word *consumer* as well as the term *producer*. The approach taken by the teacher, with reference to the term *consumer*, was to emphasise that consumers and producers have different ways of obtaining food (or energy). Consumers, for example, the animals, eat other living things for food.

Producers, on the other hand, make their own food. This approach was the normal one set out in the school scheme.

At the end of the lesson the pupils were given a test to establish which of a range of living things they considered to be consumers. Averaged over the class and questions, 77% of the exemplars which would be considered consumers by a biologist were so classified by the pupils. While there were some definite variations between exemplars (Figure 3.4) many teachers would consider such a high percentage of correct classifications to indicate that the pupils had learnt reasonably well.

However, if learning is viewed in terms of how much children's ideas have changed as a result of teaching, a different picture emerges. Prior to the lesson the pupils had been pre-tested, using a parallel test of matched exemplars and non-exemplars of consumers. Figure 3.4 also reveals that this experienced, and otherwise competent, teacher had in fact been able to make very little impression on the children's prior conceptions of consumer. After teaching, a few more pupils categorise those exemplars near the left-hand side of Figure 3.4 as a biologist would. On the other hand the opposite tendency applies to exemplars on the right hand side of Figure 3.4.

These results suggest that there must have been some problems in the way the teacher presented the concept of consumer to the pupils, or at least there are problems with what was *learnt* as a result of the teaching. In this particular lesson the new terms being introduced were related to words that the pupils were already familiar with, namely *animal*, *plant* and *living*. However, as we have already seen in this chapter, the teacher's meanings for these words may not have been the pupils' meanings and therein might have been the problem.

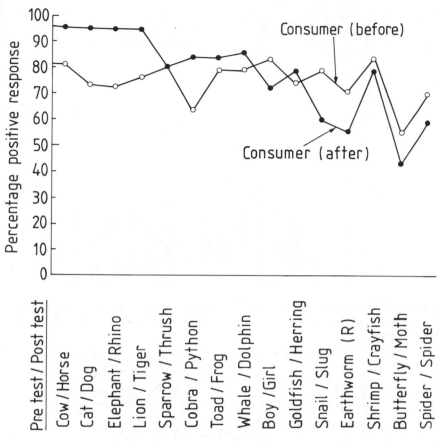

Figure 3.4: Responses before and after teaching to the question 'is it a consumer?' with a class of 13-year-olds (after Bell and Barker, 1982 — sample size = 26).

The reason for the 'strange' changes from pre-test to post-test identified in Figure 3.4 becomes clear if we consider these same pupils' ideas about animal (Figure 3.5). To be taught that animals are consumers is fine, provided pupils have a biologists' view of animal. If, however, their views about animal are those indicated in Figure 3.5, then the results of Figure 3.4 are even predictable.

What happens if we *first* teach the biologists' view of animal *before* we teach about consumers? This was done with a parallel class of pupils. The class was then taught that *consumer* was another name for the biologists' meaning for animal. The results from this class on pre- and post-tests of *animal* and *consumer* are illustrated in Figure 3.6. They clearly show that, if we can get the simple underlying words understood, then sound learning *can* occur. In this instance the problem appears to be, not with the more complex term *consumer*, but with the simpler, more common ones such as *animal* and *living*.

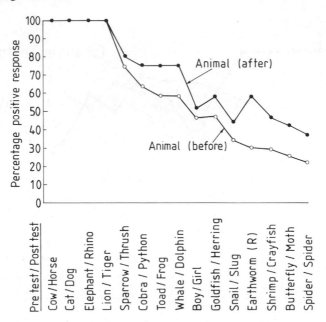

Figure 3.5: The pre-test and post-test results for the same group of 13-year-olds on 'is it an animal?' with no specific teaching of *animal* (after Bell and Barker, 1982 — class size = 26).

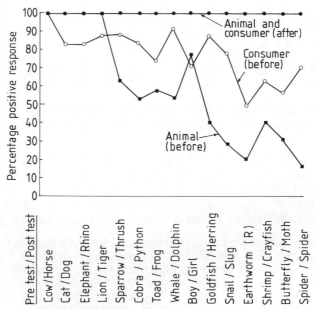

Figure 3.6: Changing 13-year-old pupil's views about 'is it an animal?' and 'is it a consumer?' (after Bell and Barker, 1982 — class size = 24).

4 Building on Children's Intuitive Ideas

Roger Osborne

Our discussion thus far has shown how some of the difficulties which children have in learning science may arise. Such problems, however, are not just a matter of semantics. Before they come near a science classroom children have already developed ideas about how and why things behave as they do — ideas which may be attached to particular terms and forms of language, but often arrived at independently of language. These 'intuitive' ideas can have a powerful influence on subsequent learning, as we shall see.

In the present chapter we will consider another content area in science teaching. Children's ideas about force and motion, about how things move and why they speed up or slow down, are frequently quite different from scientists' views on these matters. First, we will outline some of our research in this area, to provide a context in which to consider the broader issue of how to help children modify their often firmly-held ideas to accommodate those of scientists. Then, we will put forward three conditions for the success of such instruction, namely:

(i)   teaching which will help children exchange, evolve or extend their existing ideas with respect to a particular topic;
(ii)   teaching which will present new ideas so that they appear intelligible, plausible and useful to the learner;
(iii)   teaching which will order the topics of the curriculum better to take into account the learner's intuitive and/or developing ideas.

Finally, we will return to the topic of force and motion to illustrate the consequences for instructional design of attempting to take these factors into account.

Children's ideas about force and motion

Many teachers find this topic difficult to teach to younger children. For example, the following comments were made by three secondary school science teachers:

I feel I need time and resource materials from a physics-minded person to help with that particular topic. Especially at the 13-year-old level, it is trying to get it across in a simple enough way that they would understand it and enjoy it.

I wonder really how much is being attained in terms of understanding what forces are.

When it comes to teaching things like force and energy they (teachers) *get them all confused. Many teachers just throw the section* (of work) *at them in a week or two weeks, and the kids* (13-year-olds) *get rather muddled because the presentation was muddled.*

a

b

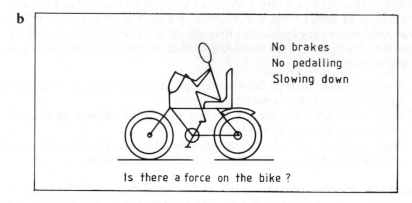

c

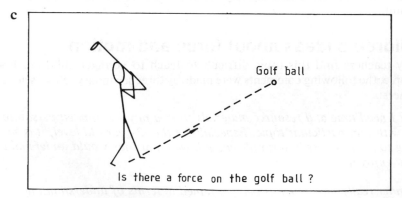

Figure 4.1: Examples of the interview cards.

Physicists would generally agree that the concept of force is a basic and central concept in the learning of physics. Some of our investigations of children's understanding of force, using the interview-about-instances procedure, have been undertaken as part of the Learning in Science Project and others at the University of Surrey (Osborne and Gilbert, 1980b; Gilbert, Watts, and Osborne, 1982). Examples of the diagrams used in these interviews are shown in Fig 4.1. At the commencement of each interview the interviewer explained that he or she was interested in the pupil's meaning for the word *force*, and that together they would 'have a chat' about some diagrams on cards. Each card was then presented, with the interviewer commenting *'Do you consider that there is a force on the_____, in your meaning of the word force?'* While a physicist would probably seek clarification of this question before answering it — whether the question was about the *total* force or *any* force, for instance — this was found not to present any real difficulties in the interview situation. The cards were simply stimuli for evoking children's ideas and if a child considered there was a need for further clarification then this was readily discussed.

Once the pupil had responded to the initial question, almost invariably with a 'yes' or a 'no', the interviewer then asked *'Why do you say that?'* or *'Can you tell me about that?'*. In this way, using 20 or so cards involving a range of situations, it was possible to build up an appreciation of the pupil's use of the word *force* and to gain a perspective on his or her views on force and motion. In the Learning in Science Project some 40 pupils aged 7-19 years were interviewed in this way.

Some pupils gave responses that were, initially, difficult for the interviewer to comprehend, even in the one-to-one interview situation. For instance, shown Figure 4.1(a) and asked whether or not there was a force on the car, a 13-year-old stated:

No, I don't think so because he is not forcing the car...the car won't move, it would be too heavy...he would be arguing at the car, kicking it... then he would probably start walking...he would do a force, if it was a brand new car...try and save his car instead of leaving it out.

This response, which has been slightly condensed, illustrates a number of features typical of other replies we obtained. The pupil emphasises the human aspect of the situation, considering that the person may be trying *to force* the car to move. But unless he or she actually gets the car moving the child obviously considers that a force is not really involved; moreover, the pupil views with some urgency the need to move the car if it is new, as one would not want to abandon it on the side of the road! We have found that these human-centred viewpoints are frequently central for children, although they would be ignored by a physicist analysing the situation. An extreme instance is that given by a 9-year-old, who stated that there was no force on the car because it couldn't feel anything!

Another quite different aspect of children's ideas about force and motion pertains to the way their ideas relate to the views of scientists in earlier times.

This can best be illustrated by considering pupil responses to Figure 4.1(b). With most of the situations there were always some young children (even aged 7 or 8 years) who had a relatively sound Newtonian perspective on force and motion, similar to that which would be taught in secondary schools today. As stated by one 9-year-old, who was asked if there was a force on the bike:

Yes. . . . because something is trying to slow it down. . . because something is pushing it the other way so it slows down.

On the other hand, some non-Newtonian but prevalent views are exemplified by the following:

Yes. . . . the wheels are still going so that there would be a force from that.
9-year-old

It is just putting force on by itself, from the force you gave it before.
11-year-old

Yes. . . . the speed that he has already got up.
19-year-old who had studied physics for two years.

There is a force because of the bike's own mass. . . the mass of the bike has come to such a speed that it won't just stop straight away. . . the force is still in there. . . in the bike. . . the force was transferred from the person pedalling. . . and it is now still adherent in the bike. . . the bike still moves forward.
20-year-old who has studied physics for three years

These all express a similar point of view: there is a force in the bike in the direction of motion which keeps it going. As one pupil said: '*If there was no force it would just stop.*'

When the golf ball situation was presented (Figure 4.1c), over half of those interviewed considered there was a force *within* the golf ball acting in the direction of motion. For example,

The force from when he hit it is still in it.
13-year-old

The force from the golf stick which slowly dies out.
15-year-old science student

There would be a force on the ball from the man hitting it. . . which would be getting less as it goes up.
16-year-old ex-science student

Yes, the force of the hit. . . gravity. . . air resistance.
17-year-old former physics student

It is interesting that this last response was made by a physics student who could identify, quite correctly, the forces that a physicist would consider to be on the ball (gravity and air resistance) but had not abandoned his intuitive idea that there was a continuing force in the direction of motion.

To examine the prevalence of the children's view, unlike that of physicists, that there is invariably a continuing force which keeps the object going, the questions shown in Figure 4.2 were designed (after Watts and Zylberstajn, 1981). The answers enable us to distinguish, we believe, between those students whose understanding of force is basically Newtonian (who will answer *a-a-a* to the three questions) and those whose view is that the force will always be in the direction of motion (who will answer *b-c-a*). The questions have been used with a representative sample of 13, 14 and 15-year-old science students (N = 200 for each year) and with 16 and 17-year-old students studying physics (N = 100 for each year). The prevalence of the *b-c-a* view was found to be 46% with 13-year-olds. This figure *increased* to 53% for 14-year-olds and further increased to 66% with 15-year-olds. At the 16 and 17-year-old level the physics students still held this view more often than any other. The Newtonian view (*a-a-a*) was chosen by less than 22% of students at all levels.

Figure 4.2: Investigating the prevalence of various views about force and motion (after Watts and Zylberstajn, 1981).

The idea that force is *in* a body, and acts in the direction of motion is a view which was widely held by 14th Century Parisian physicists headed by Buridan. Figure 4.3 shows that Buridan's view is commonly held by present-day secondary school pupils and has certainly not been influenced in the way intended by science teachers. The consequence of older physics students holding non-Newtonian views about those relatively simple situations is that their understanding of more complex notions of dynamics must involve convoluted misinterpretations of what is taught, assuming that they attempt more than the rote learning of formulae. For example, how do students with Buridan's view of force and motion interpret $F = ma$?

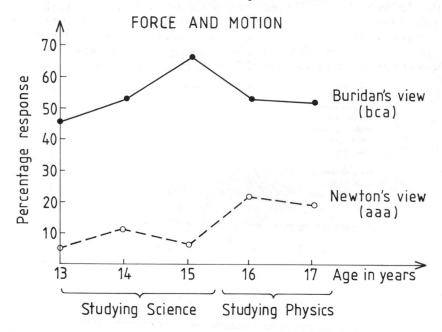

Figure 4.3: A comparison of the popularity of Buridan's and Newton's views about force and motion.

While the Newtonian view of force is a matter of definition, and in that sense simply a convention, the results presented above suggest that some of the problems associated with learning Newtonian dynamics are not just a question of semantics. Young children just *know* that you have to give an object a push to keep it moving — you put something into it. When you are moving fast, for example when you are running fast, it is hard to stop. There must be something inside you keeping you moving!

If children who have these intuitive ideas are told in a science context that there is something which affects motion called *force*, it is perhaps not surprising that they ascribe the label 'force' to the 'thing' which is put in bodies to get them moving and which makes them difficult to stop. The problem is that, in Newton's view of force and motion, this 'thing' is momentum and momentum is not considered to be a force.

Moreover, children's association of force and movement pervades their thinking about other associated concepts in physics. For example, we have found that friction is associated in many pupils' minds with wear, the generation of heat, and rubbing (Stead and Osborne, 1981a). In their thinking, two surfaces which are not in relative motion do not involve frictional forces. A box stationary on a slope is just 'stuck', in these children's view; they can thus have no concept of static friction. Similarly 'gravity' to many children *increases* with height above the earth's surface. They learn that objects dropped from a greater height cause more damage than those dropping a short distance because they 'fall faster', hence there must be more gravity acting on them (Stead and Osborne, 1981b). On the other hand, above the earth's atmosphere there is no gravity, and 'you become weightless'. On the moon there is no air and hence no gravity, although 'you can wear heavy boots to weigh you down'. Surveys suggest that some of these views are held by up to 50% of 13-year-olds.

Some general implications for instruction

The research described above is just a sample of many studies on children's views of force and motion. Other research using different investigatory procedures also tends to show the prevalence of Buridan-type ideas (e.g. Viennot, 1979; Sjoberg and Lie, 1981; Watts and Zylberstajn, 1981; Clement, 1982; McCloskey, 1983). However while a useful body of knowledge has been acquired on children's views about force and motion, as well as many similar topics, less is known about how to change those intuitive ideas — assuming, of course, that encouraging such change is an acceptable aim for science teaching (see Chapter 7).

Strauss (1981), in analysing the implications for teaching of children's intuitive ideas, suggests that the strategy to be used in any given institutional situation should depend on whether or not children already have many such ideas and how these compare to the views held by scientists. We would argue that children *always* have preliminary ideas which influence their thinking about *any* topic. In this book, as can be seen, we are particularly interested in those situations, phenomena and events about which children have firmly held ideas different from those of scientists. As teachers we would want at least some children to exchange their existing ideas for those of scientists (Hewson, 1981b), to develop a more scientific perspective (West, 1982), or to gain an additional perspective related, or relatable to, their earlier ideas (Solomon, 1983).

Posner, Strike, Hewson and Gertzog (1982) consider that, if children are to change their ideas, then they must first feel that their present ones are unsatisfactory in some way. However, dissatisfaction with a view may not in itself be a sufficient reason for discarding it. Children, like scientists, will not usually reject an idea unless they have an attractive alternative available to them. This new idea needs to be:

- intelligible, in that it appears coherent and internally consistent;
- plausible, in that it is reconcilable with other views that the child already has; and

- fruitful, in that it is preferable to the old viewpoint on the grounds of perceived elegance, parsimony and/or usefulness — although children would not put it in these terms, of course.

Moreover, children, like scientists, may come to find a view useful for a wide range of additional reasons, including how acceptable it will be to their peers if they hold that view.

Hewson (1981b) argues that any change from one viewpoint to another must be a gradual process. Ideas lose status as they become less intelligible, plausible and fruitful. Conversely, new ideas increase in status as they become *more* intelligible, *more* plausible and *more* fruitful. It is our contention that the scientists' viewpoint may often appear to children to be *less* intelligible, plausible and fruitful than their own present view (Osborne, Bell, and Gilbert, 1983), and that this is a central problem in science learning.

One problem here is that many topic areas in science teaching are considered to have a single logical order of presentation. In introductory physics, for instance, ideas about displacement and velocity are almost invariably introduced before ideas about acceleration; ideas about velocity and mass precede those about momentum; concepts in electrostatics and current electricity are taught before ideas on electromagnetism; and so on. The hierarchical order of topics employed by the curriculum developer and/or textbook writer is based on an unconscious assumption that the non-scientific notions held by pupils will not have any serious impact on the learning of conventional scientific ideas. If this assumption is wrong — as we believe it is — and if children's intuitive ideas can be ascertained, then it may well be possible to suggest alternative orders of presentation which would lead more successfully and directly to children accepting a scientific perspective.

Implications for teaching about force and motion

We have raised some broader issues which we will reconsider, and elaborate on, in later sections. In the meantime we return to the content area of much of this chapter — the topic of force — to consider specific implications for its teaching.

First, it is important that teachers become aware of children's intuitive ideas about force and motion. The survey questions in Figure 4.2 can be used to establish in any class the prevalence of the views about force, which we have discussed in this chapter. It seems desirable that teachers' guide material on this and other topics should incorporate information both on children's and on scientists' views, and in this particular instance should stress the different meanings of the term *force* and the difference between force and momentum.

Armed with this information, however, what does one do? While frictionless air tables and air tracks, and computer simulations where particles move in a friction-free environment, may help to show pupils that their present ideas are inadequate, it may well be that pupils consider these situations to be atypical, or even somewhat magical. And what is to stop children from explaining these phenomena to themselves in terms of a Buridan-type view of force and motion?

Our own attempts to modify children's viewpoints on this topic have focussed on children's intuitive ideas. We have tried first to help children to clarify and, if necessary, to re-label their ideas in a way that will help them move toward a more Newtonian perspective. Our contention has been that, if aspects of a Newtonian viewpoint can be accepted by a child as intelligible and plausible, then through thinking about phenomena in this way the new viewpoint might to found to be both elegant and parsimonious, and ultimately more useful.

In *Teaching about Force* (Schollum, Hill and Osborne, 1981) we base instruction on the knowledge that many children have a strongly held view that objects move forward because there is 'something' *in* them keeping them moving. Our first lessons, even for 11-year-olds, introduce the term *momentum* for this something. We subsequently introduce the term *force* for pushes and pulls that *act on* objects, and distinguish these clearly from momentum. We emphasise that momentum is not a force. Consideration of changes of momentum, and the introduction of ideas about the total push or pull on a body, leads to the appreciation that changes in momentum are caused by a total force acting on a body; conversely, if there is a total force on a body, then there will be changes in momentum.

In this instructional unit on force, we have emphasised the need for pupils to clarify and make explicit their own views, and to discuss these views with others. Worksheets and surveys are provided for both pupils and teachers to focus on how the pupils' ideas are changing as a result of the teaching sequence.

In New Zealand schools the concept of force is introduced between the ages of 10 and 14 years. The concept of momentum, however, is at present not usually introduced until students start to specialise in physics at age 16. Yet momentum is a scientific concept which is very much related to children's intuitive ideas (Raven, 1967-8), and we would argue that the order of introduction of the concepts *force* and *momentum* in our syllabus could well be reversed. Moreover, di Sessa (1980) has shown that many other aspects of dynamics can best be understood and appreciated in terms of momentum and momentum flow.

In our attempts to teach the concept of *momentum* to children aged 11 to 13 years we have found little difficulty. Teachers have told us that the children 'seem to have the idea already'. However, we do not want to suggest that it is always easy to teach Newtonian ideas to young children. Changes of momentum and the association of these changes with net forces are sophisticated notions. It is difficult to make the concepts plausible and intelligible, and then to show them to be fruitful.

One further point to which we draw attention on the basis of our experience, however, is that in some respects we have had more success teaching Newtonian ideas to 11-year-olds than to 14-year-olds. The older children are sometimes less interested and/or more inflexible in their ideas than younger children. While we may frequently attempt to teach ideas too early in terms of the intellectual development of our pupils, we must also consider the possibility that we may at present be introducing new ideas at a stage when children are no longer interested or do not want to be interested.

In summary, we have attempted to show how a genuine consideration of children's ideas, and the importance of these ideas to children, could dramatically change what we teach and the order in which we teach it. When such proposals are coupled with new possibilities for the presentation of information to children through computers and video disc, one can see how a sound knowledge of children's ideas could lead to a major restructuring of present-day science curricula.

5 Relating the New to the Familiar

Brendan Schollum and Roger Osborne

In the third form (age 13) *we made hokey pokey* (a type of candy[1]) *and that was really good because firstly we were having this real big thrill about making something to eat and secondly is that it was really dramatic...I mean hokey pokey making is really dramatic and thirdly is that we were all made to sit down and write up the equation of the basic thing that was happening to make it all puff up...and that was good because as we were sitting there munching away we learned something and it was fun...it was tasty and it was really interesting...we could all go home and say 'Wow...this is what we did today, Mum'.*

16-year-old girl

We would all like every science lesson to lead to this sort of enthusiasm and one might ask what it was about this lesson that made it so successful. We believe that it was because the pupil considered it relevant in a number of ways.

1   It *related* to the world outside the classroom in a way which helped the pupil expand her knowledge of that world and to make sense of it in a new way. Cooking is part of everyday life and she now knew what made the bubbles in the candy!
2   It *related* to prior ideas that she already had stored in her memory. She was able to fit the chemical symbols and equations, which the teacher had produced, into the pattern of her existing ideas and experience.
3   It *related* to what she felt someone very important in her life, her mother, would find interesting and would value.

The relevance of what children are doing in the classroom, to everyday events, to their existing ideas — and to human relationships — are all important. Pupils and ex-pupils have commented to us that whenever science lessons appeared to lack relevance their interest in science waned. So let us consider these three aspects of relevance further.

Three issues of relevance

Relevance to everyday events: Those pupils who say they lost their enthusiasm for science during their secondary schooling have frequently commented to us on what they perceived as a lack of relevance, of what they

1 Sugar, golden syrup, baking soda.

were asked to do, to their everyday lives. Some typical comments made to us by ex-pupils have been:

If we had done things in class that related to what happened everyday, that would have made it a lot more interesting.

They related it to things that you wouldn't think of, that don't really come into your life.

If I had seen more practical applications of it I would have enjoyed it more.

Teachers, too, have commented on this perceived lack of relevance. The following comments were made by very experienced secondary school teachers.

We are expecting that they will be interested in everything that we do... However quite a lot of what we do doesn't relate very directly to outside experiences... so that they can't see in a number of things where it is really fitting in for them.

We just carry out the experiments that have been set down in the books... and they often don't really relate to the outside world.

It might pay us to consider why science, which deals very much with what is happening in the 'real world', does not appear to many pupils as a subject concerned with the real world. We will return to this point later in the chapter.

Relevance to pupils' existing ideas: Relevance, for children especially, is not always just a matter of relating what they are taught to everyday phenomena. For example, consider the contrast between the following two ex-pupils. One criticised having to pull small carts along the bench top in the physics laboratory because she couldn't fit it in to anything she was interested in:

I can relate to Biology more... to things around really... I mean I couldn't relate to things like in Physics... how we push something and it hits something off the end... well O.K. So it does it... so what?

However, another ex-pupil remembering the same experience viewed it entirely differently:

The experiments we did were good... take the example of kinematics... we would take our ticker tape and cut it up and make a graph out of it... and I learnt a real lot from that because it showed you what you did... That was an incredibly good experiment... it stands out in my mind today... because I remember I learnt quite a lot from that... because we had a real good teacher.

This boy would appear to have enjoyed kinematics because it helped him understand the relationship between concepts such as distance, speed, time and acceleration in a new and more precise way. While he may or may not have been able to relate that learning to the world outside the classroom he could certainly relate the experience to his existing ideas — an experience he found challenging and satisfying.

Unfortunately all too often the scientist's view of the world does not seem to children to be related in any way to their own conceptualisations. One example of this is the particle nature of matter. Many pupils, for instance, find it inconceivable, in terms of their own experience, that air could consist entirely of particles, with nothing between the particles. Survey results from a multiple choice question about what there is between the air particles in a flask of air are shown in Figure 5.1. These have been confirmed by similar responses obtained from interviews (Osborne and Schollum, 1983) and from the work of Novick and Nussbaum (1981) in Israel. (By age 14 New Zealand pupils would have been introduced to the particle nature of matter.)

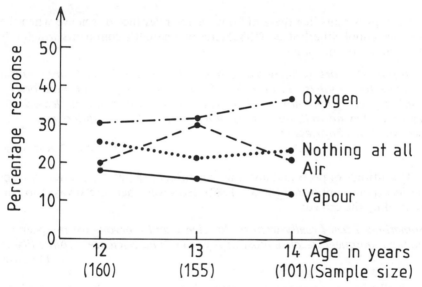

Figure 5.1: Percentage response to a question about particles (after Osborne and Schollum, 1983).

To be aware of children's existing ideas is important if we are to help children relate the ideas in their own minds to the learning experiences provided, so that sensible new ideas are constructed. We have to relate *our* teaching to *their* ideas, since we cannot control what they are thinking.

Relevance to human relationships: Pupils are undoubtedly influenced by family, friends and teachers. As Hodgson (1979) states about her own decision to study science:

The most important factor in my case, I am sure, was my home environment. The fact that I wanted to study science, far from being thought unusual or unsuitable (for a girl) *was actively encouraged.*

If parents do not show an interest in their children's work in science the child's view of the subject and its relevance to the world outside the school can only suffer further. For example:

My parents do not encourage me in science. They never ask 'How did you do in science today?' Mum doesn't ask me anything about science whereas she asks me a lot about maths because she knows I am hopeless at maths. But I have to encourage her to talk to me about science because I like science and I want her to ask me what we did that day in science.

13-year-old girl

(Do your parents encourage you to take an interest in science?) No, Mum doesn't know anything about science. (Does Dad?) No.

13-year-old girl

Perhaps parents like these did not see the relevance of science when they were at school either! Stead (1982) cites one parent's comments which refer particularly to this point.

I suppose I'm sort of fairly biased about it (science). *I don't care about it, to be perfectly honest. I don't, 'cos I was hopeless at it, and therefore I don't expect. . . you know. . . if my kids do well at it then I'm pleased at anything they do well, but if they're hopeless at it I can understand because I was hopeless at it.*

mother of a 13-year-old girl

The attitude of peers and how this can influence a pupil's attitude to school subjects is also important (Stead, 1982). However, there are always those not swayed by the crowd:

Sometimes I feel I really want to do science and everyone (of my peers) *thinks I am queer because most of my friends hate science. . . but I like it.*

14-year-old

Finally, we cannot ignore the relationships that develop between pupils and their teachers.

The most important thing for students, particularly of average ability, is their relationship with the teacher. . . Ask them what they think of science and you invariably get a comment back on what they think of the teacher.

Head of Science

To most children the science teacher is the only 'scientist' they know — certainly the only 'scientist' who knows them. When the teacher is respected, when the pupil wants to please him or her, when the pupil wants to develop a closer and stronger relationship with the teacher, where the teacher demonstrably values the pupils' embryonic scientific ideas, the relevance of science to that child can be influenced greatly.

(What are your memories of science?) Oh, it was all right. . . but then it depends on your teachers in a way. (Can you remember your first science lesson?) I was looking forward to meeting the teacher. (Were you disappointed?) No. . . when I went home and told my brother he said she was a good teacher. . . that was a big help.

ex-pupil

If relevance can enhance pupils' understanding of science, then lack of relevance can certainly hinder it. In the next section we will explore some of the reasons for feelings of lack of relevance which many children have about the scientific ideas they are taught.

The differing perspectives of scientists and children

As we mentioned earlier Kelly (1969) has suggested that we are all scientists of a sort from a young age. Children's science develops as children attempt to make sense of the world in which they live in terms of their experiences, their current knowledge and their use of language. While children are undoubtedly less conscious of their thinking processes than scientists, they make continual use of similarities and differences to organise the ideas they develop. Furthermore, in an informal and often non-explicit manner, children are continually gathering facts, developing explanations and making predictions.

However, children also bring different attributes and perspectives to the task of attempting to make sense of the world compared with adult scientists. While this might appear to be emphasising the obvious, we need to consider the nature of these differences if we are to be successful in planning and implementing suitable learning activities for our pupils in the science area.

First, and foremost, young children are limited in the extent to which they can reason in the abstract, as adult scientists commonly do. Children tend to view things from a self-centred or human-centred point of view, they tend to endow inanimate objects with the characteristics of humans and other animals, and they tend to consider only those entities and constructs that follow directly from everyday experience (Freyberg and Osborne, 1981; Gilbert, Osborne and Fensham, 1982). These features of the child's perspective contrast, as Layton (1973) points out, with the scientific perspective that has emerged in the last three centuries. Scientists constructed conceptions for which there are no directly observable instances (for example atoms, electric fields) and conceptions that have no physical reality (for example potential energy). Such concepts have certainly increased the explanatory and predictive power of science. In working out these explanations scientists have also found it useful to adopt a rather non-human-centred viewpoint, and to endow inanimate objects with human and other animal characteristics only in a metaphoric sense.

Second, children are interested in 'customised' explanations for everyday observables. They will accept more than one explanation for a specific event, and are not too concerned if some of these explanations are self-contradictory. Nor do they distinguish between explanations which might be called scientific, in that they are testable and capable of being disproved, and non-scientific explanations. And with their interest in simple pragmatic explanations for things that occur in their familiar world, children are not too concerned if two theories explaining two different situations are mutually inconsistent.

On the other hand scientists have become almost preoccupied with the business of coherence between theories. Unfortunately some of the all-embracing theories — for example, Maxwell's equations — even if they could be understood by children, do not directly and easily explain many of the everyday phenomena of interest to children. In addition, while scientists search for regularities in nature, for ways to predict events and to reduce the likelihood of the unexpected, children are frequently interested in the exact opposite, in that they wish to discover the irregular, the unpredictable and the surprise. Solomon (1980) argues that it is on this difference between children's interests and those of scientists that the discovery approach to learning science founders.

A third point to remember is that children's interests, thinking processes and their constructions of meaning are inexorably limited by their level of cognitive maturity, experiences, use of language, and their knowledge and appreciation of the experiences and ideas of others. Scientists' interests follow from the collective wisdom of the scientific community which draws upon an extremely wide range, not only of 'natural' experiences but also of contrived events and phenomena (e.g. in nuclear accelerators, spaceships, and chemical laboratories). Scientists have available to them many ways of extending the senses (e.g. electron microscopes, interferometers, telescopes), a precise use of language, and operational definitions quite unfamiliar to children. Children exposed to scientific explanations can only generate meanings from their own views of the world and their own meanings for the words used in those explanations. Appreciation of viewpoints based on our scientific history and technological culture can only develop slowly. As Gibran (1926) eloquently expressed it: 'no man can reveal to you ought but that which lies half asleep in the dawning of your knowledge'.

Finally, unless children come from a family whose members include someone with scientific training they are unlikely to interact outside school hours with people who have a scientific perspective. We have noted that young children who have a more scientific viewpoint than their peers frequently come from families with scientifically literate parents. For example:

- The only person, out of a group of 20 girls aged 8-12 years, who could successfully connect up a battery had an electrical engineer as a parent (Osborne, 1983).
- A seven-year-old boy with a Newtonian view of dynamics had a parent who was a physicist. The child had at an earlier age spent many hours playing with a frictionless air track.
- A five-year-old boy who had a biologists' view of animal, in terms of the things he classified as animal, had a parent who was a biology teacher.

Our citing of such 'precocities' is certainly not unique.

Some examples from the field of chemistry

The kids are interested in Chemistry because they are playing around with chemicals in test tubes and bunsen burners. . . in that aspect I think Chemistry is good. . . (however) it is just a game. . . I don't think they understand the concepts and some of the concepts these kids have grabbed hold of. . . it's incredible.

Head of Science, secondary school

Many children begin their formal study of science at age 11 to 13 years with high, often unrealistic expectations. Science, they frequently consider, is concerned with chemicals and explosions. The reality of the science classroom is less exciting, less dramatic, and, for some at least, less dangerous. While almost all children enjoy doing chemistry experiments, they soon find that there is more to the study of chemistry than manipulating test tubes and bunsen burners. The theoretical aspects of chemistry which the teacher introduces can produce almost immediate problems. The world of atoms, particles and molecules does not fit in with the world as they already know it. Comments made by teachers support this view.

The concept of atoms was difficult because they couldn't see it. . . sure they could see a crystal, they could see it dissolve, they could see it regrow, but what was happening to the atoms!. . . oh gosh!

chemistry teacher

Trying to convince a group of 13-year-olds kids to develop ideas of what happens inside this test tube, because chemical reactions are an important set of changes that they should be aware of, is pretty difficult. The test tube is real, and it smells too, but when you ask people to take the smell and think about it in terms of atoms being rearranged and molecules whizzing around then you have got a problem. . .

experienced science teacher

The particle nature of matter and everything that stems from that. . . it drives the kids right up the wall.

Head of Science, secondary school

Using our *interview-about-instances* and *interview-about-events* procedures, and through the observation and subsequent interviews of children involved in chemistry activities, we have begun to get some idea of the kinds of problems children face in this area of chemistry learning. Our investigations have included children's ideas about water evaporating and condensing, ice melting, candles and gas burning, sugar dissolving, water boiling, and nails rusting. Through discussion of these events with children we have been able to explore also their ideas about particles, atoms, molecules, and their interactions. This work, which is described elsewhere (Schollum, 1982; Schollum and Happs, 1982; Osborne, Cosgrove and Schollum, 1982; Osborne and Cosgrove, 1983; Osborne and Schollum 1983) has revealed some of the barriers to understanding which are faced by both pupils and their teachers.

The problem of the unobservables: Even fourteen-year-old pupils frequently have difficulty accepting that something which is not directly observable can exist or be formed (although air is normally an exception to this rule). We have found secondary pupils who believed that there was no oxygen in a test tube because they could not see anything. Many children believe that when something is burnt, such as gas in a bunsen burner, then it is just used up and nothing remains. Frequently it is assumed that air itself is not actively involved in burning although it may be considered to act as a catalyst. Many young children believe that if steam rising above a kettle is no longer visible then it must have changed *into* air. Their perception of the familiar world is different from that of scientists. It is therefore not surprising children cannot *relate* scientists' ideas to their own experience.

The problem of taste, smell and colour: Some aspects of our work indicate that children have, or develop, non-scientific ideas because of confusions about what it is that they taste, smell or see. For example, some children consider that, when sugar is dissolved in hot water, there is 'nothing left but the taste'; when a brightly coloured crystal dissolves in water then 'the colour comes out of the crystal'; when camphor is heated at the front of the room it is 'just the smell' which travels to the back of the room. Unfortunately, it is these sorts of events which are often used to introduce young pupils to the particulate nature of matter. As children's views of colour, taste and smell can be quite different from those of teachers and curriculum developers, it is not surprising that what is intended is just not seen that way by many pupils.

The problem of uninfluenced ideas: We have already illustrated, in the fields of physics and biology how many non-scientific ideas held by children appear to be scarcely influenced by science teaching. The field of chemistry is no different in this respect. For example while 37% of 12-year-old pupils in one of our representative samples considered that when a nail goes rusty it loses weight, so too did 33% of a similar sample of 16-year-old chemistry students. Figure 1.6 illustrates that 30% of a representative sample of 12-year-olds considered the large bubbles in boiling water to be bubbles of air and so did 25% of 16-year-old chemistry students. As we will discuss later in this chapter we have also found that the distribution of responses to questions about the burning of gaseous fuel were very similar for 12-year-olds and for 16-year-old chemistry students.

Even though these were not longitudinal studies we are confident on the basis of the results and interviews with older pupils, that frequently what is taught in a chemistry lesson fails to influence the way children think about familiar events. Pupils' intuitive ideas about their world can remain untouched because they are not related to the 'theoretical' ideas proposed by their teachers or their textbooks.

The problem of ideas influenced in unanticipated ways: Pupils' ideas in chemistry may be also influenced in ways never intended by teachers. While it could be argued that water evaporating from a wet plate in the kitchen is related more closely to physics than chemistry many pupils studying chemistry appear to view it as involving a chemical change. One view we found prevalent with young children (Osborne and Cosgrove, 1983) was that the water on the plate *changes* into air.

It is sort of sucked up into the air. (Is it still water?) No. (What does it change into?) Air.

12-year-old

Moreover, older pupils were able to support this intuitive view with their scientific knowledge that 'water is made up of oxygen and hydrogen', and 'air consists of oxygen and other gases'. The consequences of the juxtaposition of their earlier viewpoint and the above knowledge, is not hard to predict. For example,

(Where does the water go?) Into the air. . . it doesn't go into steam. . . it just dries up. . . (Where is it now?) Not in the steam form because it doesn't look if it has gone up in the water state. . . it must have split up because you couldn't sort of see steam or anything rising. (So when you say split up. . . what do you mean by that?) Into the hydrogen and the oxygen molecules. . . the water molecules break into their separate atoms.

15-year-old

The survey illustrated in Figure 5.2a was used to ascertain the prevalence of this oxygen-and-hydrogen view. Interviews with pupils choosing option (c) confirmed that these pupils conceived a separation of the oxygen and hydrogen in almost all cases. The results (Figure 5.2b) suggest that, if anything, the idea that on evaporation water changes into oxygen and hydrogen gases is one that becomes *increasingly* popular with pupils aged 13 to 15 years. It is also a view held by a significant proportion of 17-year-old chemistry pupils. As we have already implied a possible reason is that to the younger children the water appears to change into air — to become air. These children are then likely to learn at school that water consists of oxygen and hydrogen, while air consists of oxygen and other gases. It is therefore not surprising that children support their early idea, which is normally never challenged, with their new knowledge about oxygen and hydrogen (see Figure 5.3). Here we have a good example of how pupils might *relate* new knowledge to prior ideas in unintended ways.

The problem of models: Most young children view matter as being continuous, and this view is reinforced by what children commonly observe, for example the non-compressibility of liquids. To chemists and physicists the particulate nature of matter is a model that makes the complex world of chemical and physical changes understandable and intelligible. Yet the teaching of ideas about particles appears to bring no simplicity or elegance of explanation to a large number of children. Many of them inter-mingle continuous and particulate ideas together. As we pointed out earlier the majority of 12 to 14-year-old pupils consider that there must be something between particles (see Figure 5.1). Further, in our interview work we have found that 11-year-old to 15-year-old pupils, after observing the rapid oxidation of magnesium, produce a variety of non-scientific ideas about particles being 'shattered', 'burnt up', 'expanded', 'non-existent', 'melted', and 'shrunk'. We have found 17-year-old chemistry students still confused about the multiplicity of terms that they have been exposed to, for example particles, atoms, molecules and nuclei — and of their inter-relationships. Some of the students at this level are still unsure whether certain substances, for example steel, liquids and flames, consist of particles at all. The theoretical model or models remain unrelated to pupils' conceptions of their world.

a

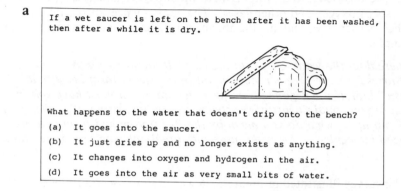

b

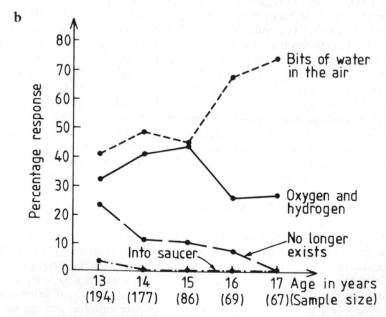

Figure 5.2: Children's ideas about the evaporation of water: (a) the survey question and (b) the results obtained (after Osborne and Cosgrove, 1983).

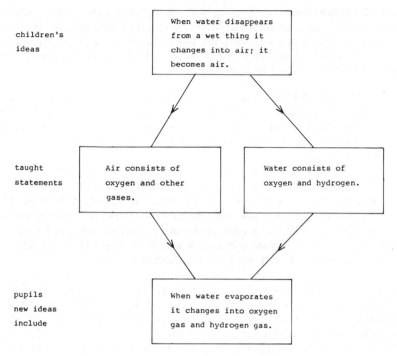

Figure 5.3: A possible explanation as to how children's ideas about evaporation can be influenced in unexpected ways.

All the above suggest that pupils can have a very real problem with some of the concepts which we introduce them to in science. Unfortunately, in many situations there is no one-to-one correspondence between the scientists' view of what is happening and what is detectable by our human senses. Yet all of us, adults and children, comprehend the world around us by placing great reliance on those human senses, and by and large, the ideas we develop on this basis are adequate in making some sense of, and predicting everyday events. Pupils bring these understandings with them to science lessons — understandings which to them are sensible and logical. It is not surprising, then, that pupils are frequently unable to relate the scientific statements they are given, either to their perceptions of the world or to their own mental constructions of how and why things behave as they do.

Towards relating the new to the familiar

What we have just said concerns the importance of children being able to relate the new ideas that they construct through one or more related learning experiences to ideas that they already hold, to other experiences and events in the world around them and to those people whose views they value. Let us consider some of the implications of this for science teaching, again using illustrations mainly from the field of chemistry.

Levels of explanation: In dealing with many topics in science, scientists themselves operate at a number of different levels of explanation. We should thus consider what level of explanation pupils can best use at a particular time and what level we wish to help pupils achieve in the long run. Johnstone (1982) refers to three levels used by chemists.

(i)   Descriptive and functional: This is the level at which chemists can see and handle materials, and describe their properties in terms of colour, hardness and so on. At this level chemists are also interested in the possibility of conversion of one material into another with consequent change in properties.

(ii)   Representational: This is the level at which chemists try to represent chemical substances by formulae and their changes by equations.

(iii)   Explanatory: This level is 'atomic and molecular, a level at which chemists attempt to explain why chemical substances behave the way they do'. Chemists invoke atoms, molecules, ions, isomers, polymers, and so on to give a mental picture by which to direct their thinking and rationalise the descriptive level mentioned earlier.

In terms of these levels there is no single scientists' view of chemical phenomena, and long before pupils are introduced to the representational and explanatory levels of chemistry it is important that they see and handle chemicals, describe their properties, and learn about changes at the descriptive and functional levels. In this way they will have something to which further learning can be related.

Familiarity: Wherever possible those materials and changes that are studied should be related, or relatable, to the familiar world outside the laboratory and also to familiar ideas in the pupils' heads. Why use strange organic chemicals to illustrate the concept of evaporation, when we could use perfume evaporating from skin and challenge pupils to generate many more examples, for example washing drying on a clothesline, the potatoes boiling dry, animals sweating, puddles evaporating, petrol evaporating from petrol spills or in engine cylinders before combustion, salt formed from the evaporation of salt water ponds and so on.

In one of the units of work which we have developed with teachers we have focussed on burning. The introduction to the unit provides information on both children's and scientists' views about burning. For example, information is provided to familiarise teachers with the results of interviews which explored pupils' views of what happens when gaseous fuel is burnt (Osborne and Schollum, 1983). While some pupils consider that 'all the gas is used up and there is nothing but hot air above the flame', others consider that 'there is still some gas left'. Other pupils have the view, like scientists, 'that different gases are formed and released'.

To illustrate the prevalence of various views, the results of a survey are provided for teachers (see Figure 5.4).

a

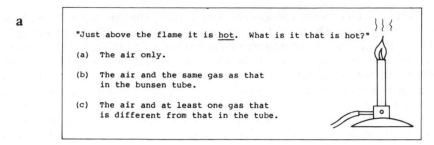

"Just above the flame it is <u>hot</u>. What is it that is hot?"

(a)   The air only.

(b)   The air and the same gas as that
       in the bunsen tube.

(c)   The air and at least one gas that
       is different from that in the tube.

b

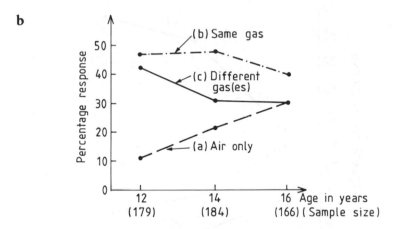

Figure 5.4: Pupils' ideas about the products from burning (after Osborne and Schollum, 1983).

Pupils could possibly choose option (a) for acceptable scientific reasons, as the products of the burning process are components of air. However, interviews with 16-year-old pupils who had chosen the option indicated that even at that level fewer than 25% had chosen it for that reason. For the remainder the gaseous fuel had been used up in the burning process leaving nothing but heat.

The suggested activities for the teaching unit on Burning draws heavily on ideas outlined by the Wreake Valley Project (Tinbergen and Thorburn, 1976). A familiar example of burning is used as a focus for the activities, the burning of a candle. The activities themselves are designed to modify children's views in a way which could help them make better sense of phenomena familiar to them. The material focusses children's attention on the yellow part of the flame, the dark inner part of the flame, and the wax, and suggests to them how they might find out more about these aspects of the burning candle. The unit then asks what happens to the wax when a candle burns? Whether air is needed? Whether air is used up in burning? And what is formed when a candle burns?

While this unit focusses on a single phenomenon, whether or not the teaching is successful has to be judged in part on the extent to which it enables pupils to develop new views, not only about the burning of a candle, but about other examples of burning as well. Following the teaching of the unit with three classes in one particular school a question about the operation of a bunsen burner was asked of pupils (Figure 5.4). The results are shown in Figure 5.5, the post-test was given immediately after the unit was taught, and the 'post-post' test refers to a surprise test administered three months later. While not all our attempts to change pupils' views have been as successful as these results indicate, at least they provide some cause for optimism as to what can be achieved taking pupil views into account.

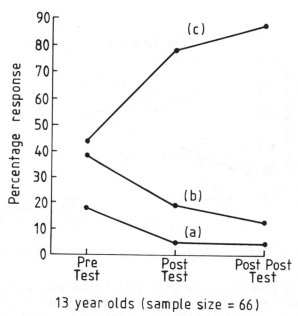

Figure 5.5: Changing children's ideas. The post-post test refers to a test given three months after the unit had been taught (Osborne, Schollum and Russell, 1982). See Figure 5.4 for key.

Readiness: In many science topics, and certainly in chemistry, it is necessary to decide at what point a certain idea or level of explanation should be introduced to pupils. An excellent example of this is provided in the case of the theory of the particle nature of matter. Our view is again that the question mainly involves the issue of relevance: the particle model should be introduced only as and when necessary, to help children make sense of their experience. In other words the question as to when to introduce a scientific idea should be answered in relation to the usefulness to pupils of such an idea. Too often, we suggest, a scientific idea is introduced because it is perceived by the teacher as something the pupils need to know rather than something which will help the pupils make better sense of the world around them. Of course this view does not preclude the proposition that teachers should so contrive classroom experiences that pupils feel the need for an explanatory model of some kind. We have to make sure that the experiences which pupils have had *before* we introduce such a model will help them see the advantages of that model and that the further experiences we organise for them will reinforce that view.

6 Facing the Mismatches in the Classroom

Ross Tasker and Peter Freyberg

Frequently, pupil involvement in a science activity is guided by written instructions. Consider the following extract from a worksheet prepared for 14-year-olds and discussed by Tasker and Lambert (1981).

Working in pairs, one pupil kneels so that his eyes are just below the level of the desk and closes both eyes. The other stands a pin upright on a sheet of cardboard on the desk top so that the kneeling pupil will be able to see the top half of the pin but not the cardboard. The kneeler now opens one eye and without hesitation brings down the tip of a ball point pen as near to the pin as possible. Measure the distance from pen-mark to pin and find the average for the class. Repeat the test with both eyes open.

For anyone who knows what this experiment is all about the purpose of the activity is clear, to show that *judgement of distance* is easier with two eyes open (binocular vision) than with one (monocular vision). The response of a class of 14-year-olds, given the above instruction, was to carry out the activity without query or hesitation; the instructions were readable and meaningful to them. A later analysis of pupils' ideas, however, showed that they had perceived the experiment in a number of quite different ways. Some thought the experiment was to do with *reaction times* 'bring the pin down without hesitation'; some thought it was to do with *time it took for the eye to focus* once it was opened; others thought it had something to do with *muscle co-ordination*. Some noticed that with their left eye open they marked the left side of the pin, and others with their left eye open marked the right side of the pin. These pupils thought the experiment was to do with *cross-eyedness*. Others, in carrying out the activity, were observed to place their hands near to the pin in readiness, obviously negating the kind of experience the teacher intended.

In each of these examples the perspective that the pupils brought to this activity was quite different from the science-oriented perspective which initiated it.

Analysing activity-based science classrooms

In many countries over the last few decades science has been taught, at least in part, by involving children in teacher-guided activity-based lessons. Through such activities children are expected to develop their investigatory skills and through the results of their experimentation to develop sound scientific knowledge. Lessons can range from teacher-paced whole class investigations to long-term pupil-paced work using instructional guides such as worksheets.

Table 6.1: Some *possible* discrepancies between teacher and pupil perceptions.

	Teacher's Viewpoint	Pupil's Viewpoint	Consequences of Discrepancy (if it does occur)
Relating to setting the scene of the lesson			
(A) Scientific context of the lesson/activity	The lesson/ activity is seen as part of a related series of experiences.	The lesson/ activity is seen as an isolated event.	Pupils do not consider the lesson/activity against important prior experiences.
(B) Scientific purpose of the lesson/activity	The purpose of the lesson/ activity is specific and obvious.	The purpose of the lesson/ activity is not obvious.	Pupils guess at and invent what seems to them to be a reasonable purpose.
(C) Scientific design of the investigatory activity	The activity has certain critical scientific features which enable particular ideas to be tested.	The critical scientific features of the activity are not recognised or understood.	Pupils are unaware of which aspects of the activity need particular attention and why.
Relating to pupils			
(D) Doing the activity	What the pupils are to do is obvious and requires skills the pupils possess.	What to do is not understood or the skills needed are not recognised.	Pupils do things which they think appropriate to reach the end of the activity.
(E) Getting results	Pupils know what to look for and will recognise the result of the activity.	What to look for and how important a result is, is not clear.	Pupils focus on unexpected aspects of the activity. They are preoccupied with 'what is supposed to happen?'
(F) Thinking about what was done and what happened	Pupils will consider critically their actions and findings.	There is no awareness that actions and findings should be critically examined.	Pupils see the activity as 'getting the right answer' or guessing what the teacher wanted from the activity.

Relating to outcomes achieved

(G) Impact of the experience on children's views	Pupils will consider the experience against the ideas and perceptions of scientists' science.	The experience is considered against ideas and perceptions of children's science.	Pupils' ideas are influenced in unanticipated ways by the experiences.
(H) Relationship to predetermined outcomes	There are predetermined outcomes which are significant for achieving the teacher's aim.	The children's outcomes of significance relate to children's ideas and children's thinking.	Pupils have difficulty relating what they did, and saw happening, to the teacher's conclusions.

Initial observation of these lessons usually shows a hive of activity, with pupils appearing to be engaged in purposeful endeavour. Closer investigation, however, often reveals a different situation. When pupils are questioned about why they are doing what they are doing, what they have found out, the significance of a particular result and its relationship to previous findings, the experimental design of the activity, or the consensus scientific view of the concept behind the investigation, their responses frequently suggest major discrepancies between the teacher's intent for the lesson and the pupils' actual involvement.

To examine this contention further, we developed within the Learning in Science Project a framework for classroom observation (Tasker and Osborne, 1983) which not only provided details of the task set as it was conceived by the teacher, but also a record of pupil action and dialogue within both group or class settings as appropriate. Tape-recorded interviews were used to obtain, from learners and teachers, insights into both their respective perceptions of the task itself and of the way the task had been carried out, as well as what had been learnt from it. The record of lessons which resulted from the transcription of interview and observation tapes, together with annotated written notes, were analysed from three perspectives (learner, teacher and observer) and in relation to various aspects of the activity, such as purpose, how it was carried out, and the perceived significance of the outcomes achieved.

Many lessons were observed in a wide range of classrooms. Sixteen of these lessons were analysed in depth. The age range of the pupils was from 11 to 14 years and nine different teachers were involved. Using a grounded theory approach (Glaser and Strauss, 1967), a set of eight issues or problems emerged, which essentially refer to mismatches or discrepancies between teacher intent or expectation and actual learner perceptions or outcomes. These are summarised in Table 6.1 and extracts and examples from the sixteen lessons will be used to elaborate each of the problem areas.

Discrepancies in intent

(A) Scientific context: Lessons were frequently seen by pupils as isolated events even though teachers saw them as building on previous lessons. For example, a teacher was involved in a lesson where the effect of heat on the rate of dissolving of potassium permanganate crystals in water at different temperatures was to be investigated. The teacher's perceptions of the context of the lesson were stated to the observer:

> *This is the third lesson in a series aimed at developing a particle idea and going over states of matter. . .the first one* (lesson) *we did was also to do with this. It was to do with expansion and contraction of substances — today was a direct follow on from that.*

The following pupil comment, however, was typical of the pupils' perceptions:

Observer: *What was today's lesson about?*

Robert : *What crystals can do.*

Observer: *Does today's work have anything to do with this other work you have been doing* (pointing to the pupil's open exercise book which was showing notes headed up as the 'dilution of a potassium permanganate crystal' and 'heating and expanding liquids')*?*

Robert : *No, not really. . .no.*

What was an obvious and strong linkage of material from the teacher's point of view was not at all obvious to this pupil. Not only would much have happened in that pupil's life since the last science lesson, but he appears to have perceived and remembered that past experience in quite a different way to that of the teacher.

(B) Scientific purpose: The purpose of the activity as perceived by pupils was frequently not the purpose intended by the teacher. This occurred with a majority of pupils in nine of the sixteen lessons analysed. There were a number of different factors which appeared to have brought this about. In four lessons, there was a failure on the part of many pupils to recognise the purpose in the sense that although the teacher's purpose was made explicit, either orally or in written form, pupils nevertheless established an alternative purpose.

Paula : *He talked about it. . .that's about all. . .*

Observer : *What have you decided it* (the task) *is about?*

Paula : *I dunno. I never really thought about it. . .just doing it — doing what it says. . .it's 8.5. . .just got to do different numbers and the next one we have to do is this* (points in text to 8.6).

On four other occasions, when the explicit purpose stated *was* recognised by pupils, discussion with the teacher revealed that this 'public' purpose was, in fact, not the actual purpose for the activity. The real purpose, was never stated and remained 'private' to the teacher.

Example
A class of 13-year-olds had established a heating curve for tapwater in a previous lesson and the teacher's intention was that they would now establish a similar curve for saltwater.

1   Teacher's 'public' purpose
Teacher   : *Right, we'll do precisely what we did yesterday.*
(to class)   *Somebody asked 'Can we put the results on the same graph that you did for homework* (the graph obtained from heating tapwater)*?' Certainly you can do that, as what we are trying to do* (today) *is compare them, isn't it?*

2   Pupils' purpose
Observer : (joining a group of three boys) *This looks interesting.*

Johannes: *Yeah, we have to. . .fill the beaker to 150. . .then um. . .get 5ml of salt, then we put it in here* (indicating the beaker of water) *and then we put it on there* (touching the gauze on the tripod) — *then we have to take the heat.*

Observer : *I see and what's it all about?*

Johannes: *I dunno — we just have to do it and get the graph.*

3   Teacher's 'private' purpose
Teacher : *We needed to use the bunsen the other day and I found*
(to   *some students had never used them, so we disgressed*
observer)   *from our programme to do something which would develop the skill of using a bunsen — also using a thermometer — and we can draw a graph. . .We did this* (same procedure) *yesterday with tapwater and someone said 'I thought water boiled at 100°' so today we are trying heating water with salt and other things in it — reinforcing the skills really.*

In this instance only instructions for carrying out the activity were given initially: the teacher eventually stated a purpose for the activity but not until pupils had completed it!

Children always try to find some purpose — even if it seems to them to be a trivial one — for what they are asked to do. If they are unable to establish a clear overall 'teacher' purpose, they substitute one of their own; such as 'just follow instructions' or 'we need to get the right answer'. We will refer to this again later.

(C)   Scientific design: Our observations have shown that pupils often did not have any idea of what were the critical scientific factors in the experiment, even though teachers assumed that they did. Pupils had little appreciation for features in the design of an investigation and consequently no real basis for anticipating the nature of its outcomes. Unfortunately, teachers we observed frequently did not sense this mismatch — at least it was very rare (only once in sixteen lessons) for them to *explain* the nature of an investigation: they gave instructions which contained design features but almost never elaborated on their significance. The effect of this lack of feeling in pupils for what they were doing was to enhance compromises in design features of an investigation and the acceptance of secondary outcomes as the significant outcomes.

For example, one of the classes we studied was engaged in a lesson on the expansion of liquids. The following information was on the blackboard.

Expansion of a liquid

Method: Fill the flask with coloured water.

Mark the level of water in the tube.

Heat the flask using a burner.

Observe what happens to the water.

Explain why this happens by talking about particles.

A number of pupils had difficulty in fitting the stopper into the flask full of water. One pupil's response was to tip some water out to make the job of fitting the stopper easier. Unfortunately this action meant that the bottom end of the glass tube was not below the surface of the water in the flask. The critical nature of this condition was not appreciated, however, and the pupils in the group started to heat the water.

The teacher noticed another group who had also not met these design requirements. She commented that they should make sure that some of the water could be seen up the tube. The pupils tried again to push the glass tube further through the stopper but when this failed they simply put the flask *without the stopper* back on the burner.

This is a clear example of how important design features of an experiment can remain obscure to at least some of the pupils in a class, to the extent that the experiment they are engaged on becomes meaningless. Worse still, not all failures to appreciate the essential conditions of a particular experiment result in such obvious defects in equipment layout or student action. The purpose of a 'control', for example, is often not made sufficiently clear to pupils. Consider the response, of two 13-year-olds, Chris and Freda, when an observer, reading in their handwritten account of a recently-completed experiment 'There was no glucose in the control tube', asked *'What's the control tube?'*

Chris   : *Was that the beaker?* (Turning to another member of the group.)

Freda   : *Another tube or something.*

Chris   : *Oh yes, we had to put in a test tube with something else inside it.*

Observer: *So this control tube is a special tube or what?*

Chris   : *No, it's just an ordinary tube.*

Observer: *Why is it called a control tube?*

Chris and
Freda   : (Shrug and giggle.)

The varied ramifications of how pupils will interpret an activity provide a real challenge for even the best of teachers.

Discrepancies in action

(D) Pupils' actions: Pupils' actions are guided by the purpose they establish for an activity. We found in our lesson analyses that, particularly where pupils had difficulty establishing any meaningful overall purpose — in a scientific sense — or where they did not appreciate the design features of the investigation, their purpose and actions degenerated to simply following instructions.

Observer: *What are you doing at the moment?*

Jim   : *Um... that one there* (points to step 3 in book) *No. 3.*

Observer: *What is that?*

Jim   : (Looks at partner who offers no comment, then picks up book.)

Observer: *Can you tell me without reading it?*

Jim   : *I've forgotten.*

Observer: *What is the whole experiment about?*

Jim   : (No response; grins)

Later the same two pupils were heating a yellow solid.

Observer: *What are you doing now?*

Keith   : *Heating this.*

Observer: *I see, what for?*

Keith   : *Well...* (races off to desk on other side of room bringing back book) *we are doing No. 5.*

Observer: *What did you do before you started heating it?*

Keith   : *These ones here* (points to No. 3 and 4 of instructions).

Observer: *Can you tell me what you have found out?*

Keith   : *We got this yellow stuff.*

Observer: *Can you tell me the purpose of this activity?*

Keith   : *No... not really.*

While 'cookbook' instructional material may be written as a fail-safe system for pupils who would otherwise not cope, there is a danger in using it that many pupils will reduce their intellectual involvement, and hence purpose and action, to a mechanical level. When this happens, many worthwhile learning opportunities are lost.

Even where the purpose and design features of an experiment are clear, there can be other problems related to pupil actions. For example, it is easy for teachers to make unfounded assumptions about the abilities and background of their pupils in terms of what is required to carry out the experimental activity satisfactorily. We have sometimes found a lack of pupil familiarity with some items of equipment and material involved in directed activities. For example, a pupil who had his own views about bimetallic strips:

Observer: *Why do you think the strip bends that way?*

Shaun   : *The metal must have been made in one way.*

Observer: *Is the strip made up of just one type of metal?*

Shaun   : *No, the handle is different.*

Some pupils talked to us about sulphur coalphate (copper sulphate) and HCI (HCl). Others were found to lack skills that teachers assumed were present, for example, the ability to cope with miniscus effects and parallax errors. Another problem we have observed has been the incorrect interpretation of seemingly clear instructions. For example, the simple instruction 'Fill a flask with water' was interpreted by a group of 13-year-olds as 'Put some water in a flask'.

(E)   Getting results: If the intended purpose of the activity is not clear and the critical design features of the experiment are not appreciated, then it is perhaps not surprising that the pupils' interpretation of what results should be looked for, and what results are important, is often different from those the teacher had expected the pupils to focus on. For example, to one pupil the significant result of heating the water, discussed earlier in **(C)**, was that 'when we heated the water it made air bubbles'. This seemed to her to be the important result.

In the situation where pupils are unable to establish a sensible scientific purpose and/or the design features of the activity are not understood, the problem can become one of guessing what the teacher wants from an activity or 'getting the right answer'. In some instances this simply leads to an abandonment of any vestige of good scientific method. In one instance we observed two pupils involved in an activity which required them to test for CO_2 being given off by a chemical reaction. The test they used required that an indicator change colour. The pupils, aged 14 years, repeated the experiment twice without any change of colour being detected. Instead of attempting the experiment for the third time they decided to ask the teacher what colour change could be expected. Armed with this knowledge they attempted the experiment for a third time without success:

Stephanie: *Oh, it is not going to work, Aileen. . . we will just pretend it changed.*

Aileen : *Yes.*

Observer : *Is that what you do when it doesn't work. . .pretend it worked. . So what do you write down? What do you think should have happened?*

Stephanie: *Yes. . . we found out nothing.*

Aileen : *Oh, it is not going to happen. . .too bad.* (gives up heating)

Observer : *What happens if you write down what actually happens?*

Aileen : (Giggle, shrug.)

Observer : *Why do you feel you have to write down what should have happened rather than what actually happened?*

Aileen : *Because it would be a waste of time. . . we would be doing it for nothing.*

Observer : *I see. If you wrote down what you got rather than what you should have got?*

Aileen : *Yeah. . .so now we know we are failures.*

Observer : *And nobody else is going to know if you write down what you should have got; is that right?*

Stephanie: *Yes.*

Aileen : *If we write down what we should have got then she'd* (the teacher) *think it's what we would have got, then we get more points!*

Observer : *Oh well, I wonder if anyone else got it* (to change colour) *or do they all do what you do?*
(both pupils smile)

(F) Consideration of findings: Not only do pupils sometimes focus on unexpected outcomes, events and episodes of an activity, but they also tend not to consider their actions and findings in any critical way. Certainly if they are concerned with simply 'getting the right answer', or guessing what the teacher wanted from the activity, they have no cause to consider their results critically in a scientific sense.

We observed an example of this when a class was required to mark the active volcanoes of the world on a map they had been given. Provided with an atlas and a list of names, pupils used the index in the back of the atlas to find where each volcano was located and then to mark it on their map. Included in the list of volcanoes was one active volcano situated 60km from the school concerned. Many pupils still used the same procedure as for unfamiliar volcanoes. Faced with the difficulty that this small volcano was not named in the atlas, they could not proceed!

Discrepancies in views of the world

Earlier chapters in this book have alerted us to one of the major problems inherent in many of the situations just discussed. Pupils' existing ideas and past experiences can result in their placing quite a different perspective on an activity from that anticipated by the teacher. This is not really so surprising, since the richness and variety of the sights, smells and so on, inherent in many activities, can lead to pupils making associations quite different from those intended by the teacher. Such associations, however, have particular significance for their impact on pupils' views, and the relationship of this impact to the teacher's predetermined outcomes.

(G) Impact of the experience: In the lessons we observed, the teachers often assumed that pupils would automatically relate the experiences of the lesson to the ideas and perspectives of scientists. Unfortunately, as we have emphasised, such experiences can be considered by pupils only in relation to their own views. When some pupils were boiling water in an activity designed to bring out ideas about change of state, for instance, they observed something quite different from that intended:

George : *When water is boiled it evaporates — turns into gases and goes off from these and turns into air.*

Koro : *The heat is sort of like. . . pushing out all these little. . . mmm. . . gas.*

Observer : *The gas is in the water?*

Koro : *I think they are in the water. . . when it gets heated. . . they come out.*

Observer : *The gas. . . has that anything to do with the water or what?*

Koro : *It is in the water.*

Observer : *It is not water?*

Koro   : *No.*

There are many other examples throughout this book illustrating the same point.

(H)   Relationship to pre-determined outcomes: It was not uncommon to see at the end of the activity the teacher write a pre-determined conclusion on the board. Often this conclusion was in typical scientific language. Sometimes the conclusions were quite inappropriate *even* if the experience provided had been interpreted in the way intended by the teacher. One teacher whom we observed, for example, heated camphor at the front of the room and asked pupils to put up their hands when they could smell it. Using the different lengths of time required for the pupils to respond, the teacher then embarked on a completely teacher-dominated 'discussion' of particles, leading up to a blackboard conclusion 'All matter is made of molecules'. *No one* could conclude this sensibly from the experience provided.

In Chapter 2 we exemplified the problem of the teacher's conclusion having little or no relationship to the ideas of the pupils (page 18). In another lesson we observed a pupil complete an experience 'to see what happens when you put a solid in liquid' (teacher's focus). The pupil had dissolved potassium permanganate crystals in water, and recorded 'his' experience by copying the teacher's blackboard notes explaining the dissolving process in terms of particles. The observer asked the pupil for his ideas about what had happened to the crystal when it was placed in water. The pupil said it had dissolved, and then elaborated:

Raymond: *As it sinks to the bottom the pressure lets out the colouring and it just dissolves on the bottom.*

Observer : *Can you tell me about this pressure?*

Raymond: *Forces it. . . like when you jump into a swimming pool. . . water pressure.*

Observer : *And when the crystal gets to the bottom?*

Raymond: *It dissolves. . . the Condy's crystal* (potassium permanganate) *goes — but not the colouring.*

The view of this pupil — that the colouring comes out and then the crystal disappears — is not an uncommon one. The teaching experience did not focus on this, however, but simply tried to impose another view without explanation or without indication of what was inadequate about the child's own view.

Implications for activity-based experiences

In this chapter we have limited our discussion to those situations where children are involved in activity-based investigations planned by the teacher to provide certain experiences, and to lead pupils to particular conclusions. We are not suggesting for one minute that all the discrepancies between teacher intentions and pupil responses, which we have illustrated in the preceding section, were found in all the lessons or with all pupils we observed. Nevertheless, we see these as frequently-found and important mismatches. They have occurred, it appears to us, because curriculum writers and teachers have tended to view pupil-investigation tasks largely from their own scientific perspective. This practice has encouraged a range of mistaken assumptions about how the learner will respond to, and what the learner will learn from, such tasks.

To reduce some of the discrepancies discussed we consider that, in any such activity-based teacher-guided lesson, there are three crucial issues which need to be faced and overcome. As teachers we must find ways of ensuring that:

- our intended purpose for the activity becomes the pupils' own purpose;
- the activity designed to achieve this purpose is understood and accepted in advance by the pupils as a sensible and straightforward method to accomplish it; and
- the pupils' conclusions are valued, discussed and related to the teacher's hoped-for conclusion.

How can we ensure that these things occur? Tasker and Lambert (1981) describe some strategies that teachers have tried successfully:

A* At the beginning of the lesson pupils are put into groups to read through the instructions for the activity to be undertaken. They discuss and report on questions or instructions provided by the teacher with the aim of ensuring that their purpose and design features are understood, accepted and appreciated: 'What are we trying to do?', 'Why do we need a control tube?', 'What does *test* mean in the sentence 'a test for starch'?'

> Pupil comment   : *This* (strategy) *helps because one person has an idea that may be quite different to yours.*

> Teacher comment: *I feel that pupils gain an overview of the activity which in the absence of such directed discussion is seldom apparent.*

B   Instructions for a particular task, written on a card, are cut up, scrambled, and placed in an envelope. Groups in the class re-assemble the material in a sensible order, discussing the reasons for the order they choose. Finally they check with the teacher before carrying out the task. The time spent by pupils on this pre-task activity is generally no more than 10 minutes.

> Teacher comment: *I found that by using this approach once or twice early in the year the pupils develop the habit and skill of considering the implications in an activity's description.*

C   Instructional material is re-considered to ensure that as little ambiguity as possible occurs. The use of short sentences, the underlining of key words, and built in questions to make pupils think about what the instructions mean, can all help. For example, the instruction:

Add the black copper oxide to acid, stirring constantly, until the black powder ceases to disappear.

can be rewritten:

Add the black copper oxide to the acid, stirring constantly. Have you added enough? Keep adding the black powder until some doesn't dissolve.

Quantities which are vague are confusing to students. For example, 'take a little...', 'put a few...', and 'add some...' give pupils an insecurity about what they should be doing. 'Put 4ml...', '...to a depth of 2 cm', and 'add 3 pieces...' all help, *provided* pupils appreciate why they are doing it.

Teacher comment: *While pupils are initially unfamiliar with precise instructions and quantified measurements, it provides pupils with opportunities to acquire basic measuring skills and develop their confidence.*

D   Pupils are given the opportunity to consider what their findings mean to them, what the findings mean to other group members, and how their group's findings stand alongside those of other groups. To do this:

(i)   Each pupil records his or her results and then considers the experience against his or her own ideas.
(ii)   Pupils put their ideas relating to the experience to other members of their group.

Discussion generates the group view.

(iii) The group view is presented to the class by a member of each group with support from any other member of the group.
(iv)   The teacher contrasts group views and directs discussion, raising issues where necessary until a consensus is reached or further activities (including reading about the presently accepted scientific view!) are decided upon. There is speculation about possible reasons for any unanticipated results.

Teacher comment: *I use this approach with all my classes now. The discussion which develops is in the language of the pupils and I can use this to work towards presenting the consensus scientific view or to direct pupils to further activities which challenge their intuitive ideas.*

Pupil comment   : *This way shows to yourself that you can understand the work that you are doing.*

E Assessment mechanisms are designed so that the things which are considered important are focussed on in the assessment procedure, and that the pupils realise this. A teacher's assessment list could include:

(i) Interpretive Skills: Ability to plan, carry out and report on an investigation.

(ii) Cognitive Skills: Ability to formulate a sensible conclusion in terms of the evidence, to interpret results and make predictions, to appreciate the views of others and the evidence on which it is based.

(iii) Manipulative Skills: Ability to set up gear, make readings, follow instructions where necessary.

(iv) Workshop Skills: Ability to tidy up after experiments, observe safety procedures.

(v) Social Skills: Responsibility to group, to class. Ability to share and listen to the views of others.

F Checklists are developed to ensure:

(i) activities are planned with close reference to the specific problems outlined in Table 6.1 (an example is given in Appendix C1);

(ii) feedback is obtained through self or colleague evaluation (an example is given in Appendix C2);

(iii) feedback is obtained through the perspectives of the pupils (an example is given in Appendix C3).

While the checklists are too detailed to be used for every lesson they can be employed in a variety of ways, for example, used formally for certain critical lessons; used informally as a guide to problems which often arise; or used in sections, perhaps focussing on one or two aspects of the checklists for a series of lessons or for a set period of time.

Teacher comment: (i)   *The planning checklist quickly exposes some of the deficiencies in the information available to pupils and helps me think more carefully about the views and skills that pupils need to have if they are to carry out the activities as expected.*

                (ii)   *I find that the self-evaluation checklist allows me to identify specific aspects of the activity that need my attention. Also a fellow teacher and I visit each other's classrooms and use the checklist to describe what we see, overhear and are told by pupils. Both of us gain a lot from this experience.*

                (iii)   *The pupil checklist has been used both with selected pupils, who have or have not interacted with the activity in the way expected, and also with the whole class. My experience has been that pupils respond objectively and provide me with worthwhile feedback for improving the learning experiences.*

Summary

In our view the keys to effective teaching in teacher-guided activity-based science lessons are as follows.

The teacher helps pupils to generate:

- a satisfactory purpose for the lesson,
- a sensible activity or method of obtaining possible solutions,
- reasonable and sensible conclusions from the experiences; and, if appropriate,
- links to, and understandings of, the acceptable scientific viewpoints on this particular phenomenon.

These keys should be on all teachers' key-rings. We have put forward some suggestions for constructing them but a central dilemma in teacher decision-making will always be with us:

How can teachers carry out what they believe to be their responsibilities when these include control of pupils' learning and encouraging pupils actively to formulate knowledge. In one direction lies control so strong that school knowledge remains alien to the learner (whether he rejects it or plays along with it); in the other direction lies a withdrawal of guidance, so that learners never need to grapple with alternative ways of thinking. The teacher has to find his way between the two.

Barnes, 1976, p. 178

Part 3

Wider Considerations

7 Assumptions about Teaching and Learning

Peter Freyberg and Roger Osborne

'The most important single factor influencing learning is what the learner already knows; ascertain this and teach him accordingly.'

Ausubel, 1968, p. iv

This statement by Ausubel is now widely accepted — possibly so because as Driver (1980) has pointed out, it can be interpreted in a range of different ways. For example,

- find the subskills that a learner has, then develop the learner sequence starting from these subskills (Gagne and White, 1978);
- find the logical structures of thought the child is capable of and match the logical demands of the curriculum to them (Shayer and Adey, 1981);
- find the prior concepts of the learner and determine the necessary links between what is to be taught to what the learner already knows (Ausubel, 1968);
- find the alternative viewpoints possessed by the child and provide material in such a way so as to encourage the child to reconsider or modify these viewpoints (Driver, 1980);
- find the meanings and concepts that the learner has generated already from his or her background, attitudes, abilities and experiences and determine ways so that the learner will generate new meanings and concepts that will be useful to him or her (Wittrock, 1974).

All these interpretations have merit but because of our particular interests we wish to focus on the views of Ausubel, Driver and Wittrock. The emphasis on the learner having to actively construct or generate meaning is one we have found especially useful in thinking about, analysing and planning science teaching and learning. It is also appealing to many people who are interested in children's ideas in science (see for example, Driver, 1982; Driver and Erickson, 1983; Pope and Gilbert, 1983).

Generative learning

Wittrock's (1974, 1977) view of learning with understanding focusses on the proposition that learners must themselves actively *construct*, or *generate*, meaning from sensory input; for example, sights, sounds, smells and so on. No one can do it for them. His perspective complements, in many ways, that of Kelly (1969) who has emphasised that man understands himself, his surroundings and his potentialities by constructing ideas about these things and by testing the usefulness of these constructions against such criteria as the successful prediction and control of events. Piaget, too, considered that knowledge is *constructed* by the individual as he or she acts on objects and people and tries to make sense of it all (see Kamii and De Vries, 1978). Knowledge is acquired not by the internalisation of some outside given but is constructed from within.

Wittrock has combined his ideas about generating meaning along with others about information processing to produce a *generative learning* model (see for example Osborne and Wittrock, 1983). According to this model:

(i)   The learner's memory store and processing strategies interact with sensory input (stimuli of the senses) available from the environment by actively selecting and attending to some inputs and ignoring other inputs. In a busy science classroom, for instance, pupils can only make sense of something if they actively select and attend to certain sights and sounds and ignore others!

(ii)   The input selected and attended to by the learner, of itself, has no inherent meaning. For example, a teacher's statement, such as 'there is force on a ball rolling down a hill', provides aural input which is simply a set of sounds. The teacher's meaning is not transferred to the learner simply by the learner hearing these words.

(iii)   The learner generates links between that input and those parts of his/her memory store considered relevant *by the learner*. Sometimes links are made to aspects of memory store not intended by the teacher. For instance, a pupil mentioned to us that the only link she was able to make on hearing the term *magnetic flux* was to events related to the soldering flux used by her father.

(iv)   The learner retrieves information from memory store, and uses this information to actively construct meaning from the sensory input. The learner cited previously may have constructed a meaning for 'the magnetic flux through the coil', but it is unlikely to have been the teacher's meaning.

(v)   The learner may test the constructed meanings against memory and sensed experience. In our example any meaning the pupil was able to construct did not relate well to other ideas in memory, nor to the classroom environment in which the ideas about magnetic flux were being discussed. The discrepancy impressed but did not help her.

(vi)   The learner may subsume constructions into memory. Sometimes new ideas may be readily accommodated alongside ideas already stored. At other times considerable restructuring of ideas and reinterpretations of experiences may be required to successfully subsume a new construction.

(vii)   The learner will place some status on the new construction, albeit subconsciously. Frequently, new constructions and previously existing ideas in memory will be held simultaneously and over time the status of one view may increase while the other decreases. As we discussed in Chapter 4, factors such as intelligibility, plausibility and usefulness, as well as a multitude of less rational and conscious thoughts, may influence this status.

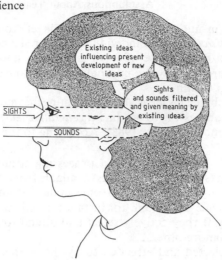

Figure 7.1: The importance of existing ideas in the learning process; a schematic representation.

In our interview work over the past few years we have frequently been surprised at the extent to which children cling to their existing views and delay accepting an alternative view in spite of the learning experiences that have been provided prior to the interviews. Yet as teachers most of us tend to assume that when we have taught something by outlining it verbally — perhaps on just a single occasion — then it will be learnt by those who have heard it or read it, *in the way we intended*.

What is, in fact, generated by a learning experience will vary tremendously from one experience to another. As the generation of a new idea is based on a person's existing ideas, plus sensory input, often only a slight modification of existing ideas need occur from the learner's point of view for the successful subsumption of the generated ideas. Attempts by teachers to rearrange the learner's existing ideas in memory in any extensive way must be fraught with particular difficulties. First there is the problem of the learner generating what the teacher intended from the sensory input provided; then there is the problem of the subsumption of the information occurring in a way that modifies ideas in memory in the way intended by the teacher. Without some appreciation of the learner's existing framework of ideas, and the extent of the modification needed for him to appreciate what is being taught, successful teaching becomes even more difficult.

In science education over the last decade there has been much discussion about developmental stages and their possible constraints on what can be learnt at each stage. A number of investigations we have reported show age-related changes in children's responses (for example, Figures 1.2, 1.6, 2.17, 3.2, 3.3, 4.3 and so on). However, we have not been able to identify any stages (Piagetian or otherwise) in such responses, except in the most general terms of decreasing egocentric and anthropomorphic explanations with age and an increasing tendency towards abstract hypothesising.

Our research has not been directed towards testing any stage theory of development, of course, but the results taken as a whole seem to indicate such a range of ages at which specific responses are given that no general stage-theory could adequately explain them. Generative learning theory in any case would suggest the alternative proposition that, within very broad developmental and ability-related limits, children will process information differently according to their experiences rather than their age.

This point of view is supported repeatedly in our interview material. Probing children's ideas and asking them to explain why they believed what they did frequently reveals a range of thinking which would not arise if children's cognitive operations reflected the generic types of structure which are the essence of a stage theory. Moreover, the range to which we are referring is far greater than could be explained by any 'transitional' phase : responses in a single interview, for instance, have ranged from (in Piaget's terms) pre-operational to formal operational.

This is not to deny a considerable similarity in the responses of individuals, and of groups of individuals, as Flavell (1982) points out. In fact, such similarities simplify, to some extent, the task of teaching. But our experience has been that learning is generally so situation-specific as to make predictions about individual responses, based upon assumed or observed stage-theorised responses, quite hazardous. Hence the need for teachers to find out what their pupils have already learnt as the result of their particular cultural and individual experiences.

Any model of learning unavoidably incorporates some gross over-simplification and no model is likely to be useful in helping us take into account all the complexities of the process. We need to keep in mind also that there are other factors involved in learning to which we have not drawn attention explicitly. For example, while new constructions might be expected to be retained if they are considered more intelligible, fruitful and plausible to the learner than his/her existing ideas, West and Pines (1983) have emphasised we cannot ignore the learner's feelings as an important component in this process. The sense of achievement, power and satisfaction, which comes from learning how to do something; the emotional satisfaction of seeing patterns in what was previously confusion; the feeling of warmth deriving from ideas and viewpoints similar to one's friends; and possibly in contrast the feeling of comfort when we consider we have found the truth in spite of what others might think — all influence the desire for conceptual change, and the very ideas we construct and accept as of value.

Whatever the limitations of the generative model it does emphasise one important aspect of learning which teachers need to consider. Any learning which requires considerable changes in the learner's existing ideas requires the learner to take a major responsibility for his/her own learning behaviour. This does not mean that learners need to organise and provide their own learning experiences, of course; this is often the appropriate responsibility of the teacher. But the generation of links between sensory input and existing ideas, and the active construction, testing and subsumption of new ideas can only be accomplished by the learner.

Finally, the generative learning model suggests to us why helping children to learn is so difficult. To contrive a learning environment such that pupils are highly likely to generate the meanings teachers would want them to achieve, and to modify their existing ideas in hoped for ways in the process of subsuming these new ideas, requires considerable teacher skill, knowledge of pupils' present ideas, and an understanding of their ways of processing information. As Nussbaum and Novick (1981a) have suggested, the more we appreciate the complexities of the learning process the more amazing it becomes that we can teach anybody anything!

Assumptions about the teaching-learning process

In the preceding chapters we have referred to many specific aspects of children's science and to the widespread persistence of their intuitive ideas despite formal teaching. We have also commented on various general attributes of those ideas including how children tend to have self-centred or human-centred viewpoints, how their views are based on their everyday experiences and common use of language, how they are interested in specific explanations rather that coherent theories, how they tend to endow inanimate objects with the characteristics of humans and other animals, and how non-observables are considered not to exist. We have also seen how children can endow an object with a certain amount of a physical quantity (for example, force) and for this quantity to be given an unwarranted physical reality. Related to this is the problem children have with models (for example, the particle nature of matter).

If one adopts a generative view of learning — that new ideas can be constructed only through generating meaning for sensory input by using existing ideas in memory store — the importance for science teachers of understanding children's science becomes very clear. It should lead teachers to adopt in their planning what Fensham (1980) calls a 'student dominance' assumption. This recognises that children's science must necessarily play a dominant part in science learning. Moreover, since children's ideas are based on their learning over their entire life-time, it is not surprising that these ideas are felt by children to be quite consistent with one another and are thus unlikely to be easily modified or exchanged for an alternative viewpoint, through a few lessons on a science topic.

Unfortunately the student dominance assumption, upon which teaching could be based, has tended to be overlooked in the development of science curricula. In fact the existence of children's science has usually been ignored or at best inadequately considered. It is extremely rare to find curriculum guide material which explains to teachers the likely views of children in their classrooms, and ways in which these views could most readily be modified. Rather the implication which underlies many science curricula is that the learner possesses no significant knowledge of a topic, or any knowledge which could have any bearing on the learning, prior to the teaching episode. This is the 'tabula-rasa' assumption; the learner has a blank mind which can be filled by the teacher.

An alternative assumption, upon which present teaching is also frequently based, is that the learners have some prior ideas related to a topic which is going to be taught, but that these ideas can be directly and easily replaced. This 'teacher dominance' assumption is based on the view that the teacher's ideas can be somehow transferred to the learner and old ideas will be quickly discarded as 'inadequate' or 'incorrect'.

Our view is that both the 'blank-mind' assumption and the 'teacher-dominance' assumption have been the root cause of many of the problems which we discussed earlier. On the basis of what is now known about children's science Gilbert, Osborne and Fensham (1982) described a range of possible outcomes from teacher-learner interactions on children's science, which take more account of the inevitability of 'student dominance'.

1   The undisturbed children's science outcome.   We have already cited many examples of this. While older students may include more sophisticated or technical words in their descriptions, their views appear to remain essentially those of much younger children. For example, consider the following non-scientifically-acceptable views about gravity.

The higher up you go, the stronger the gravity is until you get out of the atmosphere.

14-year-old

The higher up you are the stronger gravity is because if you jump down from something high up, you're obviously going to fall a lot heavier than if you jumped from something lower down.

20-year-old

2   The two-perspective outcome.   Sometimes students find it advantageous to construct, and store in memory, aspects of the scientists' viewpoint, even though this is not really the way they tend to think about things. While this may be a desirable situation, and even a necessary one in the often gradual process of accepting a new idea as more intelligible, plausible and useful than an old idea, often the two viewpoints remain unrelated. One view is for 'living', the other is for examinations; and no attempt is made to perceive, or reduce, contradictions inherent in holding the two opposing viewpoints. For instance, we have found 17-year-old physics students who, despite the fact that they had learnt and could recall the inverse square law of gravitation, still held to the idea that gravity increased with height above the earth's surface (Stead and Osborne, 1981b).

3   The reinforced outcome.   Sometimes ideas that are taught are unintentionally misinterpreted as lending support to the learner's present ideas. As an example in Chapter 5 we discussed how learners often consider that on boiling, or evaporation, water changes into air. Their view is reinforced by learning that water contains oxygen and hydrogen, and air contains oxygen and other gases. A similar kind of instance appears in a textbook which states 'Where does friction occur? Frictional forces come into operation when two moving surfaces come into contact' (Petchell, 1976, page 100). The commonly held view that frictional forces occur only between surfaces in relative motion (Stead and Osborne, 1981a) seems likely to be reinforced in the case of pupils using this text.

4   The confused outcome.   Sometimes learners, through their learning experiences, lose confidence in their earlier ideas despite these ideas being relatively coherent and sensible to them prior to a teaching episode. If they lose confidence in their existing ideas without better ones being constructed, however, learners may be reduced to a state of confusion and incoherence which is helpful neither to their self esteem nor to their future learning. For example, a student-teacher who had studied physics at the 16-year-old level was shown a diagram of a person pushing a car along the road and asked if there would be a force on the car in her meaning of the word. Her reply was *'yes. . . kinetic. . . it is moving from the person into the car. . . the previous one* (a picture of a person pushing a car but the car not moving) *is potential, isn't it?. . . potential energy '*. Such muddled ideas in long-term memory store are not helpful to this person, nor, worse still, to her future pupils!

5   The unified scientific outcome.   The aim of science education is to enable learners to make better sense of their world by helping them restructure their ideas in useful and useable ways. Many pupils, despite the problems of science learning which we have outlined above, do build up coherent scientific perspectives which they can relate to other aspects of what they learn, and to the world in which they live. Let us exemplify this from a 15-year-old-boy's view of gravity:

It's a force exerted by bodies and the further the distance, the less the force — all objects have it and the bigger they are the more gravity they have.

This range of outcomes of science teaching is perhaps not remarkable in terms of a generative view of learning. Children can already hold what is, to them, a perfectly satisfactory view or explanation. In constructing meanings in science lessons links can be made to ideas in long-term memory which have been developed within, and are only loosely related to, science lessons; links do not necessarily occur between the intuitive ideas which children hold about how and why things behave as they do and ideas generated in the science classroom. Furthermore, meanings can be generated in science lessons which are perfectly compatible with other ideas in long-term memory, but which are vastly different from those intended by textbook writers or teachers. Should we be surprised, then, if children are sometimes unwilling or unable to undertake the necessary major restructuring of ideas in long-term memory necessary to accommodate the ideas put forward by scientists?

The place of science in the school curriculum

During the course of investigations we have frequently asked the same questions of adults as we have of children, although not so systematically. The existence of the 'two cultures' to which Snow (1961) directed attention in the 1950's, seems to us as demonstrable today as it was then. When we discuss our questions and answers with many adults, well-educated though they may be in a non-science culture, they often ask us why we want to change children's ideas, especially as they (the adults) have got on well enough with similar notions all their lives.

Such questions lead us to broader questions about the place of science in the school curriculum and the goals of science education. If children's ideas are important in learning; if generative learning is a useful way to view learning in science; if children's ideas are influenced in a variety of ways by present science teaching; if the assumptions upon which present curricula are designed are inadequate; then what are the implications of our research for the aims of science teaching?

Firstly we point out that children will construct meanings for themselves as they attempt to make sense of their interactions with the natural and technological worlds they live in, whether we like it or not; they are involved in some kind of learning pertaining to science, both in and out of school, every day of their lives. In this respect teachers are teaching science whenever they help children:

(i)   to investigate things and explore ideas;

(ii)   to ask useful and productive questions;

(iii)   to seek and develop explanations that are sensible and useful to them, with respect to the natural and technological worlds that they confront daily;

(iv)   to broaden their experience of nature and technology; and

(v)   to become interested in the explanations of others about how and why things behave as they do and in how such explanations have been obtained.

While all these can and, in our view, should occur at all levels of education, it is more difficult to decide how much and at what stage children should be introduced to the widely accepted scientific viewpoint on a certain topic, or even be strongly guided toward that perspective. If a scientific proposition is too sophisticated for children, in relation to their prior experience and level of cognitive maturity, then it will not be appreciated; worse still, its obtuseness (for a child) could lead to a reaction against learning science and even against science itself. On the other hand, as we have pointed out in Chapter 4, it is very difficult to change certain basic ideas with older pupils, for example, ideas about force and motion. The introduction of the seed of a scientifically acceptable idea, or a nudge of the child while quite young toward a scientific perspective, may be just what will help that child in the long run. The problem, however, is to decide just when a pupil will be able to benefit from visual and verbal input which deliberately encourages the scientifically acceptable viewpoint.

How keen we are to introduce this viewpoint, to whom, and when, again relates to what we perceive to be the important goals of science education. In a scientific and technologically-advanced society it is desirable that all citizens should have a positive attitude to inquiry and to developing their own ideas. In addition it seems to us desirable that as many citizens as possible understand something of the viewpoints of scientists, appreciate how scientists arrive at those views and be able to recognise some of the limitations implicit in them. Furthermore we want some of our young citizens to become the scientists and technologists of the future. This leads us to suggest that the aim of science education for children should be to ensure that they are *all* encouraged:

(i)   to continue to investigate things and explore how and why things behave as they do, and

(ii)   to continue to develop explanations that are sensible and useful to them.

For the future of our scientific and technological society, however, our goals must include more than this, at least for many children. Our view would be that without jeopardising (i) and (ii) we would want *many* children:

(iii) to recognise that scientists have sensible and useful ways of investigating things, many aspects of which apply not just to science, and

(iv)   to regard at least some scientific explanations as intelligible and plausible and as potentially useful to society, if not to the child personally.

Also, where it is possible to do so without jeopardising (i) to (iv), we would want *some* children:

(v)   to replace their own intuitive explanations with, or to evolve their own ideas towards, the accepted explanations of the scientific community, and

(vi)   to become committed to the endeavours of advancing scientific knowledge still further.

If these aims are accepted, then for some children at least, and for some topics, we need to promote conceptual change in a particular direction — both for the sake of the child's own future learning in science, and as a precursor to achieving the broader goals of a changing society. But if we wish to avoid alienating many pupils from science, we must take care not to insist upon conceptual change at the expense of children's self confidence, their enthusiasm and curiosity about the world, and their feeling for what constitutes a sensible explanation.

8 Roles for the Science Teacher

Roger Osborne and Peter Freyberg

Up to this point we have been mainly concerned with the importance, for teachers, of taking children's ideas into account; the reasons for this; our view of the learning process; some teacher assumptions that can undercut teaching effectiveness; and the consequences of all these for science teaching. We have also raised questions about language and communication, about relevance, about sequence in the curriculum and about science activities which are not achieving their purpose. We will now consider some implications these matters hold for the science teacher and some of the more successful teaching strategies which they might suggest for such teachers.

All teachers have to assume many roles, but teachers of science have, if anything, a wider range of roles to perform than those of some other subjects. The organisation and maintenance of laboratory equipment, the development of safe routines for the handling of potentially hazardous materials, the diverse nature of the activities engaged in by pupils both in the laboratory and in a more formal classroom setting — all these make special pedagogical demands on the science teacher. Moreover, any approach to teaching which aims to take into account the ideas of children, and the processes by which they construct new ideas, will not reduce the number of roles for the teacher but will rather enhance the importance of some existing roles as well as generating at least one new one (Nussbaum and Novick, 1981a).

We will consider especially the roles of motivator, diagnostician, guide, innovator, experimenter and researcher.

The teacher as motivator

One widely accepted role of any teacher is that of pupil motivator. Often the problem in the science laboratory is not one of getting pupils to attend, even to the equipment or to the activity taking place, but to help them to attend to the 'right' things. Any activity-based science lesson involves interesting events which may be viewed, literally, in many different ways. The focus that the teacher intends is not always the one adopted by all pupils.

Osborne and Wittrock (1983) have suggested ways in which we can capitalise on the learner's voluntary control over his or her attention. The more interesting the possibilities inherent in an activity, the more important it is that attention be focussed on the 'right' things:

(i) Explicitly state the intent of the lesson or activity, so that pupils can reconstruct for themselves the problems to be solved or the learning task. For example, ensure that written material has clearly-worded headings, sub-headings and focus questions.

(ii)   Encourage pupils to ask themselves, and each other, questions which will focus attention and initiate generative learning, e.g. What does that suggest to you? What could we use that for? Why do you think that happened? What is another way we could do this?

(iii)   Encourage pupils to take responsibility for, and to direct, their own learning. Deliberately reduce the cookbook aspects of instructional material to a minimum, leaving a greater opportunity for pupils to make as many decisions about their work as possible — *provided*, of course, that the teacher acts as a 'backstop', querying what decisions have been made and why.

(iv)   Choose situations of demonstrable interest to the pupils wherever possible. Young children are fascinated by the unpredictable or unexpected event in an otherwise familiar world. As long as such events are not considered to be simply a trick or magic, they provide a challenge to children, focus their attention and generate further interest.

(v)   Encourage pupils to reflect on their own ideas and on the ideas of others. Our experience has been that pupils of all ages are very interested in other children's ideas, even quite abstract ideas such as the various views of electric current held by pupils (Figure 2.16). Introducing pupils to the range of views held by others allows them to clarify their own thinking in a non-threatening environment.

The teacher as diagnostician

This could be regarded as a new role by many science teachers. However, if children's already-existing ideas have a major influence on learning, then it is essential for the teacher to be at least sensitive to his or her pupil's ideas. If teachers are aware of some possible views held by children at various age levels, then they can devise appropriate ways to ascertain, through the use of questionnaires, informal discussions or interviews the particular views held by *their* pupils. As a medical practitioner diagnoses the cause of a symptom before attempting to alleviate it, so the teacher needs to diagnose the viewpoints of her pupils before deciding how to set about modifying them towards more scientifically-acceptable ones. Where pupils' views are completely unknown, an awareness of the significance of pupils' views can in itself lead to the discovery of some important factors in children's present thinking about the topic concerned. For this to occur regularly, however, a systematic recording of interesting comments made by pupils will need to have become second nature to the teacher.

It is usually not possible, in ordinary class interactions, to explore any one pupil's ideas in depth. However, small changes in emphasis by the teacher can assist here. For example, when an inappropriate or unexpected answer is provided by a pupil in a teacher-led discussion, a few moments can be spent attempting to find out *why* the pupil gave that answer. So often in class the inappropriate answer is ignored and the teacher moves the question on to another pupil in relentless pursuit of the 'right' answer.

To discover or to diagnose children's existing knowledge we must thus provide plenty of opportunities for pupils to express their ideas, whether in small groups or in whole class settings. However, this in itself is not enough. As teachers we also need to ensure a classroom climate where children's ideas are valued and listened to. The role of teacher as listener is inherent in the role of 'teacher as diagnostician'.

Another frequently-overlooked method of diagnosing the framework of ideas within which our pupils are operating is to analyse not only their incorrect responses to questions in class, but also their responses to test and examination questions, laboratory accounts, and problem-solving attempts. If we are to gain further insights into these cognitive frameworks we need to take every available opportunity to talk to pupils about the reasoning behind their responses to formal assignments.

The teacher as guide ·

There are various kinds of guides. For instance, an Alpine Guide approach to teaching has been described by Ogborn (1977), with the teacher leading pupils through a reasoned argument or towards the solution of a problem using a step-by-step series of questions. Each step in the exploration is turned into a question, the answer to which provides the basis for the next step. A question which is not answered, or is answered incorrectly, is regarded as inadequate and the guide attempts to replace it with a less difficult, clearer or simpler question.

But there are pitfalls in this procedure, as in most mountain climbing; while the teacher may feel elated when the top of the mountain is reached, the pupils themselves may feel no real sense of achievement. When the teacher does not clearly indicate the goal of the exercise, nor explain the purpose of the discussion, neither the path that has been followed nor the view from the mountain top is appreciated. Certainly this approach does not encourage pupils to attempt to climb their own mountains nor provide them with useful skills for doing it. As a teaching strategy, it fails by a large measure to acknowledge that pupils must learn how to construct their own paths. Directing pupils up more gentle slopes, with encouragement from behind, may be a more useful and appropriate technique in the long run. How, then should a teacher attempt to fulfil the role of guide? The prime requirement is to help pupils develop strategies for the effective processing of information, so that they can both see where they are going and have some idea of how to get there. We can:

- gently point out logical errors in a person's thinking, such as inconsistencies or the drawing of unjustifiable inferences;
- challenge the reluctance of some pupils to consider all possibilities or to suspend judgement; and
- show pupils where they have over- or under-generalised, or based their argument on false assumptions.

Opportunities for taking false steps abound in most science lessons. Often we are in a hurry to get across the facts, rather than taking time to develop sound thinking procedures which will help pupils make better sense of the world about them.

All pupils need guidance in linking their current experience to existing ideas in long-term memory as they attempt to generate meaning. Helping pupils to relate what is being taught to appropriate propositions, episodes and images already in memory store, and to think about relevant past experiences, is all important. The conception of learning as a generative process also suggests that the teacher as guide should be providing pupils with many examples and applications of a new idea, should present material in several different ways and formats (e.g. diagrammatically as well as verbally), and should encourage further elaboration of a new idea through a consideration of it from a number of points of view (for example, historical anecdotes, technological applications, mathematical formulations, societal implications and/or philosophical considerations, where these can be given at a level appropriate to the child). Pupils can be encouraged to construct headings, tables, pictures, alternative explanations, flow charts, inferences and summaries as they absorb information. They can be encouraged to make and check predictions based on their 'new' cognitive constructions. They need the opportunity to act upon the information they encounter, not just to passively receive it.

The guidance and help necessary for pupils to learn generatively also requires an active teacher continually interacting with individuals and groups as they learn, and consciously promoting generative learning. We have yet to find any written instructional material that can replace the teacher in this role.

The teacher as innovator

Teachers have had to provide the material resources for creating an effective learning environment. The supply of books, equipment and expendables (for example, chemicals, paper tape, film), together with the organisation of field trips and vicarious experiences (for example, films and other audio visual support), are all part of the science teacher's job. In addition, however, the teacher should be seen by pupils as a human resource — a source of ideas on how to do things, about where to find things, about what could have gone wrong and about what to do now. Providing 'new' or alternative ways of doing things is very much part of these responsibilities.

We can also think of the teacher, however, as an innovator at another level. Possibly the most challenging task of all, once children's ideas are known, is to find new ways of helping them to perceive the ideas of others (for example, scientists) as more intelligible, plausible and potentially useful to themselves than the ones that they presently hold. Even when children's ideas are well known and documented, and where the accepted scientific view is clear and unambiguous, it is not at all obvious how best to go about modifying their existing ideas. Knowing more about the preconceptions of their own classes, teachers are in a potentially better position than textbook writers — including the authors of this volume — to devise specific ways of modifying pupils' ideas on a particular topic. Using one's imagination in teaching stimulates teachers as well as children.

The teacher as experimenter

Teacher education is sometimes considered to be just a means for providing teachers with a kit of skills required to teach children. Since these skills are neither simple in themselves nor directly teachable, however, it is more useful to view teaching as an activity which relies, for any improvements, individual or profession-wide, on a continuing component of experimentation throughout each teacher's career.

No matter how experienced we may be, we are never completely successful in our endeavours and it is always appropriate to ask 'How could I have taught more effectively?' This requires teachers to evaluate systematically what they have done, including how much pupils have learnt as a result of particular teaching episodes.

The testing of pupils at the end of a set of lessons is frequently used to establish how much they then know about the topic which has been taught. The results of such tests should not be confused, however, with how much pupils have learnt *as a result of the lessons*, as we demonstrated earlier in our description of the experiment on teaching about consumers (Figure 3.4). If teachers are to take their role of 'teacher as experimenter' seriously it is clearly inadequate for them to assess pupils only at the end of a teaching sequence. Not only are pre-lesson and post-lesson comparisons essential, but a comparison of these results with an assessment of achievement some months after the teaching is also desirable. Figure 5.5 illustrated how pupils' views moved further toward the accepted scientific view over the months following teaching. Figure 8.1, on the other hand, exemplifies a dramatic change in viewpoint as a result of teaching but with some regression toward children's science over the next few months.

Recognising that children are likely to have sensible and coherent views which are nevertheless different from those that are scientifically accepted, we need then to consider what form the assessment of learning should take and what the feedback consequences on pupils of that assessment are likely to be. There is little doubt that the formal methods of assessment often used in science teaching at present simply encourage pupils to rote learn and pay lip service to the accepted scientific viewpoint. Usually no account is taken of the extent to which they really believe that what they learn in science lessons is important for understanding the real world. Perhaps we should be more concerned to assess:

(i)   the coherence of pupils' own views and their reasons for holding those views;

(ii)   whether or not pupils understand the accepted scientific perspective; and

(iii)   what attempts they have made to relate the two viewpoints where these differ.

To illustrate the problem further consider the following comments of an 11-year-old pupil some months after a set of lessons on electric current. (The corresponding teaching sequence is described in Chapter 10.) In the interview Vanessa quickly and successfully connected up a simple battery-bulb circuit

a

> Consider what some people say about the wax when a candle burns.
>
> In your view, which is the best statement about the wax?
>
> (a) The wax is burned up in the candle flame.
>
> (b) The wax is not burned up; it holds up the wick as the wick burns.
>
> (c) The wax is not burned up; it melts and stops the wick from burning too fast.

b

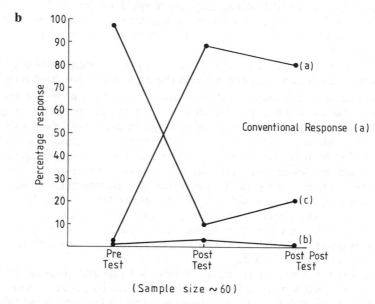

Figure 8.1: Changes in viewpoint of 13-year-old pupils as a result of teaching, together with subsequent changes over a 3 month period; **a** the question, and **b** the results (Osborne, Schollum and Russell, 1982).

when asked to do so. She was reminded of the four models (Figure 2.16) and asked which one she thought was best. The transcript has been abbreviated slightly.

Vanessa   : *At first I thought it was Model B, because I didn't realise that if it was B the battery would go flat very soon. So now I think Model C is best...quite a bit of current comes into the bulb, and some of it wasted up. It can't take it all at once so some of it takes it back.*

Interviewer: *Did you do things with meters?* (Experimentally test which Model is valid)

Vanessa   : *Yeah, we found that Model D worked with the meters. But I don't understand D. I know it works but I don't understand how it works.*

Interviewer: *Did your teacher discuss it?*

Vanessa   : *Yes, and I kind of knew how that went. . . it made sense
a little bit. . . but afterwards I thought of some other
ideas. . . How did that work?. . . What happened? and,
How did he explain. . . ? I got muddled up. If you are
not using any power in the bulb how is it going on? But
I understand C.*

How are we to assess Vanessa's learning? Now that we know what she is
thinking it would be harsh, even unreasonable, to say that she has not achieved
anything, just because she failed to choose Model D as the correct model to
illustrate flow of electric current in a simple circuit.

She deserves credit for her attempts to work through the problem even
though we must not fall into the trap of praising any answer irrespective of
its basis, or of its correspondence with the accepted scientific view.

Take a second example. Susan, aged 12 years, has been studying the topic
of force, using the momentum approach mentioned in Chapter 4. Some
months after the lessons she was interviewed about her ideas at that point. In
response to a question about the force on a golf ball (Figure 4.1c) she was
unsure whether her earlier intuitive view about the 'force of the hit in the golf
ball' was appropriate.

Interviewer: *Is there a force on the golf ball?*

Susan   : *Gravity pulling it downwards — and the force of the
hit. . . No, there is not the force of the
hit. . . because. . . Oh yes, the force of the hit. . . and. . .*

Interviewer: *So which way is that one* (the force of the hit)?

Susan   : *That was* (gives the direction of the ball's flight).

Interviewer: *Yes. Are there any other forces?*

Susan   : *I can't think of any more.*

Later in the interview the above discussion was picked up again.

Interviewer: *Can I talk to you again about the force of the hit* (on
the golfball), *because you said to me. . . there is the
force of the hit on it. . . no, there isn't the force of the
hit on it. . . yes, there is a force of the hit. Can you tell
me what you were thinking about?*

Susan   : *I think the teacher told me that the force of the hit
stops the instant the golf club doesn't touch the ball.*

Interviewer: *After it has been hit?*

Susan   : *Yes, I think.*

Interviewer: *You think... but that is not the way you thought about it before?*

Susan   : *No.*

Interviewer: *You like to think of the force of the hit* (being) *in the ball?*

Susan   : *Yes.*

Both the pupils in these two examples accept that there is another perspective to the one they have themselves, but they still find it more sensible to think about the situation in their own way, at least for the present. Somehow, through teacher experimentation, we need to develop new assessment techniques and reporting procedures which will allow both the teachers' and the pupils' intellectual integrity to remain intact, but which in addition will encourage pupils to reconstruct their ideas in such a way that they are more scientifically useful as well as coherent.

Finding out what children know prior to and after teaching has taken place is one way to evaluate how successful a teaching sequence has been. Equally important, however, is some evaluation of the nature and quality of the learning experience itself. Here the help of a fellow teacher can be invaluable. Through monitoring the nature and quality of pupil-pupil interactions using an audio-tape and/or video tape recorder, and by having an observer talking to pupils as the activities progress, a teacher can obtain feedback which is otherwise largely inaccessible. In our experience it is difficult for teachers to monitor their own classrooms at anything other than the level of 'do they appear to be getting on with their work?' It is possible to answer 'yes' to this question and yet for there to be no generative learning occurring in that classroom. Pupils can switch off and yet go through the motions, as Barnes points out.

Pupils can form and even play the system, but many do not allow the knowledge presented to them to make any deep impact on their view of reality.

Barnes, 1976

Adopting a generative learning view, the science teacher as experimenter should be *centrally concerned* with the impact of teaching on his or her pupils' view of reality.

The teacher as researcher

Teaching can be a lonely occupation. Often teachers can complete almost their entire careers (except for their initial training) without really having observed another person teach. Specialist science teachers at adjacent schools (for example, physics teachers) can teach for years without meeting one another. While a teacher can be an inventor and even an experimenter in such 'splendid isolation', he or she cannot fulfil the role of the teacher-researcher without contact with other teachers. To be an effective researcher in most fields it is important to have the opportunity to report one's findings to other researchers and to have their comments. Much of the most successful experimentation in teaching involves sharing findings about children's ideas and ways of thinking, and sharing in discussions about the successes and failures of innovative teaching strategies and sequences. Unfortunately it is all too easy for the role of teacher-as-researcher to be confused with that of teacher-as-marketeer. A teacher with a 'new' teaching package or idea can be so keen to 'sell' it to others that a cold and dispassionate analysis of the context in which it has been used, its weaknesses as well as its strengths, and its limitations as well as its potential, are not given sufficient attention.

On the other hand, where teachers are able to become effective researchers sharing their findings with other teachers, the level of professionalism is very high. Despite the frequent reporting of failure in some, if not many, aspects of their aspirations for a teaching sequence, morale can remain buoyant (Freyberg and Osborne, 1982). We are convinced that many teachers have this potential for obtaining and sharing findings of a research nature — not world-shattering findings perhaps but by no means insignificant — which would add spice to their teaching and make it more creative and satisfying.

Whether or not a teacher undertakes research into learning, what is so important is that a teacher does have enthusiasm for enquiry in general. We have been fortunate, over the years, to have been invited to observe science and other lessons in classrooms in Britain, the United States and Australia, as well as New Zealand. Some of those lessons, from the observer's point of view, have been highly successful, others less so. What has stood out, in our minds, about the successes we have seen is the way in which the teacher concerned conveyed to his or her pupils a sense of continuing involvement in exploring some small corners of the world, as though there was always the possibility of something new and interesting turning up. It was their enthusiasm for enquiry which was so contagious, and which we regard as one of the most important goals of science teaching.

In this chapter we have considered the contribution of the roles of motivator, diagnostician, guide, innovator, experimenter and researcher to the process of generative learning. Recognition of the importance of such roles, we believe, will help teachers develop those particular teaching strategies necessary if significant learning is to be achieved. In the next two chapters we will discuss some teaching sequences which incorporate these roles for the teacher.

Part 4

Towards A Teaching Model

9 Lesson Frameworks for Changing Children's Ideas

Mark Cosgrove and Roger Osborne

In Chapter 7 we advanced the merits of adopting a particular perspective on learning — the generative model — and then went on to suggest some general strategies which teachers could adopt to promote such learning. What we propose to do now is to consider how the generative learning model can assist in the planning of a specific lesson or set of lessons designed to modify children's intuitive ideas. In doing so we face a problem of our own making : to be useful, our discussion needs to propose particular sequences of classroom activities, yet our theoretical perspective suggests that these sequences cannot be planned until we have found out just what it is that the class already knows. Hopefully we will be able to suggest ways in which teachers can exploit the gap between the horns of this particular dilemma.

Several frameworks or models for the planning of science lessons designed to change children's views, and based on similar views of learning to our own, have been·proposed in recent years. Possible templates for science teaching sequences have been suggested by Barnes (1976), Karplus (1977), Erickson (1979), Nussbaum and Novick (1981b, 1982), Renner (1982) and Rowell and Dawson (1983). Building on their work, and incorporating the findings from our own studies, we now put forward our own views on both typical teaching sequences and the specific teaching strategies which should be associated with them. These need to be considered in relation to the more general strategies, or teacher roles, discussed in Chapter 8. Finally we will discuss some of the problems involved in providing teachers' guide material which not only assists teachers to select the set of activities that they undertake with pupils, but which also encourages — for better or for worse — the adoption of specific classroom teaching strategies. A specific example of the application of the teaching model to a lesson series on electric current will be given in the following chapter.

A plethora of models

Let us first review some of the teaching sequences that have been proposed over the past few years in an endeavour to help children develop more scientist-like ideas. As we have already demonstrated, present-day science teaching has at best been only partially successful in achieving this aim.

According to Renner (1982) the most common practice of science teachers is to attempt to pass on to their pupils a mastery over content as this content is envisaged by the teacher. The theory of learning underlying this approach is, first, that the material to be taught can be given to the learner as *information*; second, that it should then be *verified* by the learner through observation; and finally, that the information should be *applied* in some way to 'settle it in'. The first stage, informing or telling, is usually attempted through the teacher's opening statements or as the introduction to a so-called 'experiment'. This subsequent activity is not an experiment in the investigatory sense, however, but merely a verification or demonstration since both pupils and the teacher already know the expected outcome. The final stage, in this typical science sequence — application of knowledge — usually involves answering questions and solving quantitative or mathematical problems from a textbook in preparation for a test of some kind.

Renner's analogy for this entire process is that of a guided tour where the guide, the teacher, points out all the sights to be observed and the learner is discouraged from taking any detour that, in the guide's view, is not productive. While the track may become familiar to the learner, it is often excessively narrow and paved with the veneer of learning which may be very thin indeed (Fensham, 1984), particularly where the teacher's assumptions about pupils' prior ideas on the topic are false.

If we accept that each of us must develop the understandings we have about a concept for ourselves, then Renner suggests an alternative teaching model as more appropriate. His initial concern is with pupils gaining experience and this becomes the first stage of his teaching model. Learners are provided with suitable experiences in order to create for themselves what is to be learnt. In the second stage, the learner is introduced to some appropriately-specific terminology in relation to the phenomenon being investigated. The teacher uses this to assist the learner to interpret what has been found. In the third stage, the new ideas of the learner are meshed with existing knowledge in order to expand both that knowledge and the newly acquired idea. Additional experiences to help this elaboration process are an essential part of this stage. These experiences would have some of the attributes of experiments because the *outcomes* would not be known even though the pupils know the *concept* that is the subject of investigation.

In summary, Renner's view is that much conventional science teaching is simply a training process which involves telling, confirming and practising. Its limitations are obvious. From the generative learning point of view, it omits the vital activities which involve originating experiences, interpretation and elaboration.

Some other models which have been proposed also reflect a similar viewpoint to Renner's in that they demonstrate a real concern for the cognitive development of the learner. One such model was proposed by Karplus (1977) and he, like Renner, has been somewhat influenced by the Piagetian theories of development. Karplus argues that science learning should be a process of self-regulation in which the learner forms new reasoning patterns. These will result from reflection, after the pupil interacts with phenomena and with the ideas of others. Karplus also proposes a three-phase learning cycle. The first phase is one of *exploration* in which pupils learn through their own actions and reactions with minimal guidance, while the teacher anticipates few specific accomplishments. The learners are expected to raise questions that they cannot answer with their present ideas or reasoning patterns. In the second phase of the Karplus model, the concept is introduced and explained. Here the teacher is more active, and learning is achieved by *explanation*. Finally, in the *application* phase, the concept is applied to new situations and its range of applicability is extended. Learning is achieved by repetition and practice so that new ideas and ways of thinking have time to stabilise. An interesting analysis of the use of this particular learning cycle with one topic is provided by Smith and Lott (1983).

A similar three-stage model (see Figure 9.1) has been suggested by Nussbaum and Novick (1981b, 1982). They sought to explain what happens as learners change their conceptions during instruction. Their strategy, in common with all of the models summarised here, is based on the principle that 'science concept learning involves cognitive accommodation to an initially-held alternative framework'. Or, as we would prefer to put it, the teaching task is to ascertain individual children's conceptions about science topics and to modify these *towards* the current scientific view. To bring about cognitive accommodation, Nussbaum and Novick suggest that the first step is to *expose the alternative frameworks*.

They note Ausubel's warning that 'preconceptions are amazingly tenacious and resistant to extinction' (Ausubel, 1968), and accept that such preconceptions often interfere with the teacher's learning outcomes. Thus, Nussbaum and Novick propose that the first step in facilitating accommodation should be to ensure that every student is aware of his/her own

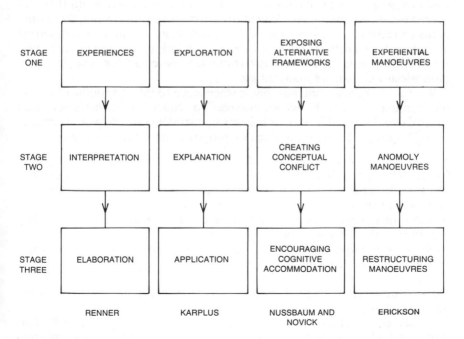

Figure 9.1: Three-stage teaching sequences proposed by Renner (1982), Karplus (1977), Nussbaum and Novick (1981b, 1982) and Erickson (1979).

preconceptions. To them, this is most easily achieved if some event can be devised which requires learners to make explicit their existing ideas in order to interpret it. Pupils are encouraged to describe their own views verbally and pictorially, and the teacher assists them to state these ideas clearly, in order to recognise what they can and cannot explain. Pupils are encouraged to debate the various views represented by all of their fellow learners, in order to better understand the features of each view.

Assuming that learner dissatisfaction with their existing ideas results from such activities, and the teacher provides additional experiences which will lead to further dissatisfaction, then conceptual conflict is likely to result. Nussbaum and Novick imply that this conflict must be sufficient to induce students to recognise that their existing views require modification. Accommodation develops from pupils searching for a solution to their conflicting ideas. Hence, in the Nussbaum and Novick model, concept learning is achieved by *exposing* alternative frameworks, *creating* conceptual conflict and *encouraging* cognitive accommodation.

Erickson (1979) makes a parallel set of proposals. The first stage of his model is the provision of a *set of experiential manoeuvres*, which allow the learners to become familiar with a *wide* range of phenomena, so that they might expose a set of intuitive ideas or beliefs. In this stage, the activities are considered in sufficient depth to allow the learners to clarify their ideas and to develop confidence so that they may begin to make predictions. The second stage contains *anomaly manoeuvres*, involving the creation of situations that lead to unexpected outcomes. An element of uncertainty is introduced; the learner needs to restructure his/her views. The third stage is a set of *restructuring manoeuvres* to assist the learners in accommodating unexpected outcomes. Restructuring, in Ericksen's strategy, could be achieved by, for example, group discussions and teacher intervention.

Barnes (1976) also contends that learners need to take a prominent part in the formulation of their own knowledge. To reduce the teacher's perceived control over knowledge, Barnes believes that students should work primarily in small groups. In practical terms he proposes the following sequence:

• *a focussing stage*, in which the teacher, with the students, prepares the ground by presenting preliminary knowledge (which, we assume, includes 'alternative frameworks' and 'children's science'). When the attention of the class is fully focussed on the topic, the teacher moves on to an

• *exploratory stage*, involving much discussion and other activities, including experimentation. Then in the

• *reorganising stage*, the teacher re-focusses attention and tells the groups how they will be reporting back, and how long they have to prepare for it. Finally, in the

• *public stage*, the groups of learners present their findings to one another, and this leads to further discussion.

A further model has been suggested by Rowell and Dawson (1983). This more explicitly focusses on the confrontation between children's science and scientists' science. They suggest the following sequence:

- Through questioning, the teacher establishes the ideas which children bring to the problem situation. Conscious awareness of these ideas is of value to both the teacher and the children.
- These ideas are accepted by the teacher as possible solutions.
- Children are asked to retain their ideas, and the teacher states that he or she is going to put forward another possibility which the children will help in evaluating later.
- The 'new' idea is taught by linking it to a basic idea already held.
- Once the new idea is available to children then the old ideas are recalled for comparison, with each other and with reality.

Rowell and Dawson believe that children are less threatened by this approach than some others, since both 'old' and 'new' ideas are the pupils' own in the sense that all are pooled knowledge. Assuming that old theories are rarely defeated by contrary evidence but only by better theories, they argue that the children with several ideas available to them are in the best possible situation to accept the scientist's one when it is tested against the others. On the other hand, Solomon (1983) argues that where children's old ideas involve 'everyday notions' and new ideas involve 'scientists' explanations, pupils are likely to retain both kinds of 'knowledge'; they need to be able to think and operate in both everyday and scientific domains, but they should also be able to distinguish between them. We contend that while this may be true for some concepts, (for example, animal, plant, living), it may not be valid for others (for example, force, electric current), where everyday notions are neither necessary for everyday communication nor useful for explanatory purposes once the scientific perspective is fully appreciated and understood.

Yet another model?

We have now briefly outlined several models proposed by others for changing children's ideas. In many ways they are more remarkable for their similarities rather than for their differences. Why then propose yet another model?

From our own work on the Learning in Science Project, we have come to recognise a number of preconditions associated with teaching and learning, which we consider should be taken into account by teachers if they are to be successful in modifying some of the firmly held ideas that pupils usually have already. None of the models just described, in our view, pay sufficient attention to these inter-related conditions. They are:

First, the teacher needs to understand the *scientists' views, the children's views*, and *his or her own views* in relation to the topic. In many situations there is likely to be some discrepancy between the teacher's views and those of scientists. For example, Shipstone (1982) studied the widespread persistence of fallacious models of electric current, and noted that of a group of eighteen *physics* graduates undertaking a postgraduate course in teacher education only eleven of the group were firmly committed to a Model D (Figure 2.16) view of electric current — that electric current is conserved in a circuit. In a similar group of graduate biologists and mathematicians, 45% used unscientific models.

We would argue that curriculum developers need to put considerable effort into clarifying and analysing the range of views children hold, and the scientific views, on the topic in question. Teachers' guide material then needs to provide teachers with the information on children's views, and contrast these with scientist's views. Survey material designed to establish what views a person holds can be used by a teacher first to check on his or her own views, and then to establish the prevalence of the different views within the class. Time needs to be allocated in the teaching schedule for this latter activity.

A second precondition to successful science learning is the opportunity for pupils to explore the context of the concept, preferably within a real (everyday) situation — for example, the construction of simple electrical circuits can be the context for exploring the concept of electric current. Usually we cannot afford to let this be only a brief encounter, particularly where there is some aspect of the context which is unfamiliar to the prior experiences of pupils. Contexts which can be related by pupils to everyday and non-laboratory experiences, and to ideas of social significance, help pupils link the experience to relevant ideas and may help motivation. Inagaki and Hatano (1983) argue that the acquisition of knowledge is initiated either by a natural curiosity to make sense of the world or from the desire to induce an event which people value in some sense. They believe that when these two motives are combined, the initiation of information-seeking will be accelerated. When something that is expected and desired does not occur, individuals are motivated to ask 'why' and to seek further information. In their view teachers can best motivate learning by redirecting such 'why' questions into the form 'how can we make it happen?'.

A third precondition is for the learners to engage in a self-clarification of their own views at an early stage of the teaching. A major factor in many learning situations is the learner appreciating what ideas he or she already has which are relevant to the topic, and how these ideas differ from those of other people. An important task for the teacher is therefore to provide early opportunities for learners to clarify their existing ideas on a topic within a supportive environment, so that they do not feel threatened by revealing what they may feel to be queer or hazy ideas.

Next, learners need opportunities to debate the pros and cons of their current views with each other (Nussbaum and Novick, 1981b; Barnes, 1976). This procedure almost inevitably captures their attention (Osborne and Wittrock, 1983) since they feel called to defend their own ideas, are led into a sustained personal interest in the discussion, and concentrate on the issues of importance and relevance to them. We contend, as well, that the act of presenting one's own views and considering their merits in the light of other views is a useful simulation of the behaviour of scientists (and others who are interested in a pursuit of knowledge and understanding).

A practical aspect of this classroom debate is whether or not it should occur in a whole class or small group setting. Barnes (1976) has argued strongly for the small group situation. Gilbert and Pope (1982) note, however, that pupils need opportunity to develop certain skills if all group members are going to have the opportunity to articulate their views. They have to learn, for instance, to refrain from dominating a group discussion. They also need to learn how to bring the silent group members into the discussion, how to avoid blocking out members by ridicule and other strategies, and how to suppress their own recognition-seeking behaviour. The monitoring and guidance of a number of groups which are simultaneously engaged in discussion in the one classroom, requires considerable skill as well as physical and mental agility on the part of the teacher. The passing contributions of the teacher to the group discussions, however, should provide a lead to other group members as to the probing techniques and group handling processes they should adopt.

At the conclusion of a discussion, a wide range of possible views should have been introduced, mostly by pupils themselves but occasionally by the teacher. As Rowell and Dawson (1983) emphasise, it may be the responsibility of the teacher to bring up the scientists' view or views, as an alternative to those put forward by the learners. Sometimes this alternative view, because of its abstractness, its structure or because it is counter-intuitive to pupils, is not received with much enthusiasm until it can be rendered intelligible and plausible by experimentation, demonstration or reference to analogy. This is a crucial stage in affecting a change of status between the differing viewpoints.

Even in situations where it is possible to provide a critical experimental test which invalidates all but one of the views proposed, some learners will not necessarily change their views at this stage. They still find their own view satisfactory because it links more coherently to other ideas they hold. Neither do they always accept their teacher's interpretation of the experimental design and outcomes provided; for example, they may consider the evidence to be a trick or a special case. And, as we have discussed earlier, there may be other reasons, including less rational and coherent ones, why scientific views consistent with the evidence are not immediately accepted. Teachers must sometimes accept this as a fact of life, and decide to leave the learner's integrity intact. It can be argued that it is better for the learners to remain honest about which view they prefer and why, rather than to accept the teacher's view superficially. Unfortunately most current examinations and test practices will not reinforce this view and where the teacher is also the examiner, some pupils will naturally be reluctant to discuss their understandings openly.

There is one further precondition to successful science teaching, in our view. If the teaching sequence merely presents the scientific concept with supporting evidence and does not provide opportunity for the consolidation and elaboration of that idea, it will be surprising if some pupils at least do not revert to their earlier conceptions. They have to be given the opportunity to consider the new concept in a range of situations, or across a range of exemplars. Classroom activities which enable, or require, learners to use their new concepts to explain specifically observed events are of particular value; the new view is established as useful and the status of that view is likely to be increased. We have found it especially useful for such purposes to provide a realistic set of practical problems based on domestic and other technological situations.

To sum up, the model which we thus propose for changing children's ideas reflects our desire to achieve three specific objectives: *clarification* of the pupils' existing views, *modification* of these views towards the current scientific view, and *consolidation* of the scientific view within the background experience and values of the pupils.

The 'generative learning' model of teaching

The teaching sequence we propose, and its associated teacher strategies, can be considered a teaching model for *generative learning* in that it is based on the generative (constructivist) view as well as attempting to meet the preconditions discussed in the last section. In common with many of the teaching models we have reviewed, it has three distinctive teaching phases (*focus, challenge,* and *application*), but these are preceded by an explicit teacher preparation phase (Table 9.1).

Preliminary phase
We have argued that teacher preparation should include:
- ascertaining the typical ideas which children will bring to the topic, together with the prevalence of these views in the class concerned;
- understanding the ideas that scientists use to describe and explain these phenomena;
- an open appreciation by the teacher of the ideas he or she tends to use to describe and explain the phenomena.

The teacher preparation we are talking about here requires for most teachers a particular type of guide material which will include survey questions that the teacher can use with the class. For some topics the historical development of a scientific idea may be a useful introduction for teachers. Sometimes the early ideas held by the scientific community resemble children's views and, when teachers and pupils realise this, it can encourage more open discussion about, and respect for, children's ideas in the first two phases of the teaching sequence. An understanding of the critical historical experiments which led to scientists changing their ideas may also have implications as to how children's views might be changed.

Where research data and hence teachers' guide material is not available to help teachers in the preparation phase, a specific class exercise some weeks before a topic is to be taught, attempting to extract pupils' ideas through discussion or written responses, may yield information helpful for planning the set of lessons.

The focus phase
The aim here is to provide a context for later work. This could involve activities to focus attention on particular phenomena (for example, the relationship between pushes and pulls and motion), or to get pupils thinking about their own meanings for words (for example, the consideration of a set of interview-about-instances type cards), or a combination of both. The teacher's role consists of providing motivating experiences, encouraging thinking through the questioning of pupils about what they think, and helping pupils to interpret their responses. The purpose of this phase needs to be made explicit to pupils if they are to take responsibility for their own learning; to become familiar

with the context, to ask questions of themselves and others, and genuinely to attempt to clarify their own views.

Table 9.1: A teaching sequence based on the generative learning model.

Phase	Teacher Activity	Pupil Activity
Preliminary	Ascertains pupils' views; classifies these; seeks scientific views; identifies historical views; considers evidence which led to abandoning old views.	Completes surveys, or other activities, designed to pin-point existing ideas.
Focus	Establishes a context. Provides motivating experiences.	Becomes familiar with the materials used to explore the concept.
	Joins in, asks open-ended personally-oriented questions.	Thinks about what is happening, asks questions related to the concept.
	Interprets pupil responses.	Decides and describes what he/she knows about the events, using class and home inputs.
		Clarifies own view on the concept.
	Interprets and elucidates pupils' views.	Presents own view to (a) group (b) class, through discussion and display.
Challenge	Facilitates exchange of views. Ensures all views are considered. Keeps discussion open.	Considers the view of (a) another pupil (b) all other pupils in class, seeking merits and defects.
	Suggests demonstrative procedures, if necessary.	Tests the validity of views by seeking evidence.
	Presents the evidence for the scientists' view.	Compares the scientists' view with class's view.
	Accepts the tentative nature of pupils' reaction to the new view.	

Application	Contrives problems which are most simply and elegantly solved using the accepted scientific view.	
	Assists pupils to clarify the new view, asking that it be used in describing all solutions.	Solves practical problems using the concept as a basis.
	Ensures students can verbally describe solutions to problems.	Presents solutions to others in class.
	Teacher joins in, stimulates, and contributes to discussion on solutions.	Discusses and debates the merits of solutions; critically evaluates these solutions.
	Helps in solving advanced problems; suggests places where help might be sought.	Suggests further problems arising from the solutions presented.

The challenge phase

At this point, pupils can present their own views to a group or the whole class. The differing views held by members of the class are sought, displayed and discussed. Where necessary the teacher introduces the scientists' view at an appropriate level of sophistication and language for the pupils. Critical tests for checking on the various views are proposed preferably by the pupils, where applicable, and evidence for the scientists' views may be sought. This phase of the learning is a critical one in terms of teacher input and guidance, and as to what is practical in the classroom. It may require considerable whole-class teaching and discussion. Since small groups may need constant teacher input for their maintenance, control and outcomes, it would be impractical for the teacher to be with all groups at the same time. If the challenge phase is successful, then, it should end with pupils raising many questions as they try to accommodate new ideas. In fact the success of the phase can be assessed largely on this criterion.

The application phase

For many topics this can be a problem solving time where ideally the solutions require the accepted scientific viewpoint, or can be achieved more simply through the use of that view. Discussion about the proposed methods of solution should enhance the process of problem solving and increase the status of the new ideas. Again an active teacher role is essential — diagnosing pupils' existing ideas, encouraging pupils to attempt alternative solutions, challenging pupils to think about phenomena in terms of the new viewpoint, and encouraging a reflective thinking approach by the pupils.

Encouraging classroom implementation

Effective implementation of the ideas we propose requires active teaching by a teacher who clearly appreciates children's ideas, the scientific view to be encouraged, the types of activities which might achieve conceptual change and the associated interactive teaching sequences which need to be adopted. Generative learning requires active teaching and in our experience no material written directly for pupils can replace the important continuing interactions between teacher and pupils.

How can teachers be oriented to this different view of teaching and encouraged to adopt teaching sequences and strategies which will persuade pupils to change their ideas? In pre-service and in-service courses for teachers it is possible to introduce children's ideas on a particular topic, then the scientist's perspective, and then introduce a possible teaching sequence and associated teacher strategies through the use of a video-tape of model lessons or excerpts from a sequence of lessons.

Where formal courses for teachers are impractical reliance must be placed on teachers' guide material. The form which this material takes is crucial. If the teachers' guide material does not detail specific classroom activities only the most capable teachers are able to utilise the general ideas which we have been expounding about obtaining children's views and scientists' views and developing these within a series of classroom activities. On the other hand many teachers require more specific guidelines. Without them they are not prepared to proceed. Nevertheless such material can be too detailed : either it will not be read or if it contains instructional material for pupils it may be reproduced directly for pupils in the mistaken belief that the materials will do the teaching.

In our view, teachers' guide material for the average teacher must set out a relatively clear framework for classroom activities. Wherever possible guidelines for teacher roles and strategies need to be suggested *alongside* details of the proposed lesson activities. Teachers who are unsure of themselves can follow the suggested activity sequence, but also are continually made aware of the specific strategies that can and should be adopted to ensure, and monitor, that worthwhile learning is occurring. The more confident teacher will in any case develop her own programme.

10 A Teaching Sequence on Electric Current

Mark Cosgrove and Roger Osborne

In Chapter 2 we considered children's ideas about electric current and some of the problems that can occur in teaching elementary ideas about electrical circuits. Hence this topic provides a useful context in which to exemplify the lesson framework discussed in the previous chapter, and to review the teaching strategies that we have discussed.

The teaching sequence and teachers' guide material on electric current which we now describe were developed in an attempt to build upon:

(i)   our own and others' findings about the ideas on electric current held by children and by more senior physics students;

(ii)   our conviction that, without a basically sound appreciation of the scientist's view of electric current in a simple circuit, pupils are likely to seriously misinterpret other terms associated with electrical circuits, e.g. resistance and potential difference;

(iii)   our interest in exploring the use of technological knowledge and situations as a context in which children could reconsider their existing knowledge as well as evaluate new ideas and ways of thinking about what happens in electrical circuits.

Electric current provides a good example of our suggested teaching strategies and sequences because:

(i)   children have, or quickly acquire through experience with electrical systems, relatively firm ideas about 'electricity' flow in circuits. *Electric current* is a frequently used term in our technological society and it is one that pupils quickly adopt as a label for their existing ideas.

(ii)   It is possible to devise a critical experiment to show that only one of the various ways in which pupils think about electric current in a simple circuit is consistent with experimental evidence.

The teaching sequence we eventually fixed upon was based on a set of investigations involving both interviews and classroom observations. These included our own studies on children's ideas about electric current (for example, Osborne and Gilbert, 1980; Osborne, 1981) and on changing children's ideas about electric current in a simple circuit using a one-to-one pupil-teacher situation (Osborne, 1983). This work formed the basis of a teaching plan for the overall sequence, but particularly for its 'challenge' phase. The sequence was then trialled with several groups of children and these efforts were video-taped (Osborne, Tasker and Schollum, 1982). Use of this video-tape material with teachers, and a reconsideration of the sequence in the light of the various pre-conditions for learning discussed in Chapter 9, led to the production of the final programme and teachers' guide material (Cosgrove and Osborne, 1983). We now describe this present sequence.

The preliminary phase

The Teachers' Guide on Electric Current (Cosgrove and Osborne, 1983) first provides a two page introduction to the distinction between children's science and scientists' science. It is also suggested that, as teachers, we need to be aware of our own views, and a warning is given that changing children's ideas towards the scientists' view is more difficult than most people imagine.

Material is then provided in the guide to help teachers clarify their own views about electric current. Interview-about-instances cards, used originally by Osborne and Gilbert (1980), are illustrated. These provide specific situations for teachers to consider. Responses which might have been made by a physicist, but which were in fact made by a competent student, are included for each situation to indicate the physicist's view (Figure 10.1). Finally the range of views on current held by pupils are presented to the teacher.

The preliminary phase thus provides the teacher with the physicists' view and children's views, and helps the teacher clarify his or her own ideas on the topic. This information may also suggest to teachers ways in which they might survey or quiz members of their own classes about the existing views held by its members.

The focus phase

In the Teachers' Guide we set out a number of experiences for pupils involving simple electrical circuits. Typically the circuits have not more than one or two cells, one or two lamps, a switch and some insulated hook-up wires. The suggested activities are provided to enable pupils with little or no background knowledge to make up their deficiencies, and for pupils who have had previous experience to extend their experiences. The suggested activities are summarised in Table 10.1.

We designed the activities in this phase so that pupils could work through them at their own pace. Once they are engaged in the activities, however, the teacher needs to interact continually with individual pupils, or small groups of pupils who are working together. As the pupils construct and use simple circuits, the teacher increasingly seeks to *focus* their thinking on the electrical happenings in their circuits. Towards the end of this phase, questions like 'What do you think is happening in the wires (or lamps)?' serve this purpose. However, any question raised by the teacher needs to become the pupil's question for meaningful learning to occur subsequently.

The challenge phase

This is the crucial phase of the teaching sequence where the pupils' own views about electric current are clarified; they learn about the views of other pupils and their teacher; they reconsider their own views; they design, or are introduced to, a critical test of the various views; and they are confronted with the test evidence.

Sample cards and questions

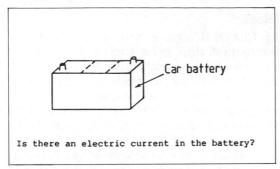

Is there an electric current in the battery?

A physicist's view

'No, I don't think so...there may be some minor currents that are not supposed to be there. The battery is a store of electrical energy but there is no flow of charge.'

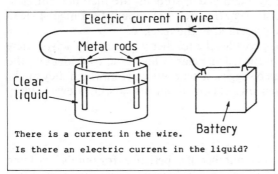

There is a current in the wire.
Is there an electric current in the liquid?

'If there is a current in the long wire there must be an equal current in all parts of the circuit. The liquid must be a conductor otherwise there would be no current in the wire. In the short wire the direction of the electric current is from rod to battery.'

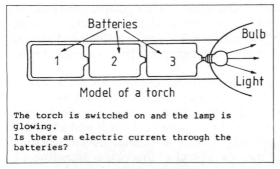

The torch is switched on and the lamp is glowing.
Is there an electric current through the batteries?

'Yes, all batteries have the same current through them as they are all connected into the same circuit. The currents in all parts of the circuit will be the same.'

(supplementary question where applicable)

Which battery has the most current through it?

Figure 10.1: Examples of the interview-about-instances situations provided in the Teachers Guide (Cosgrove and Osborne, 1983).

---

* **Making the lamp glow**. Use a lamp-holder, lamp, two plastic-insulated wires, and a 1.5 volt cell. Pupils suggest and learn ways of testing whether or not the lamp is defective or whether the cell has expired.

* **Introducing the Morse Code**. Useful in itself, the Morse Code also provides a number of opportunities to make circuits and to encode, transmit and translate simple messages. Small-group cohesion is developed in an atmosphere best described as play. Pupils find out about telegraphy, and become involved in the process of learning a new language.

* **Extending the signalling equipment**. Pupils suggest ways of using the circuit with two switches, one for each operator, two lamps, and two cells, so that messages can be sent from one room to another. Pupils are invited to record any questions they might have about electrical effects. Questions can be written on file cards and fixed on to a pin-board.

* **Keeping records of circuits**. The teacher asks pupils to make accurate drawings of the circuits they construct. This soon becomes a frustrating exercise because the drawings take 15-20 minutes to complete, and pupils are usually dissatisfied with their efforts. They can try to describe their circuits in words, but this also becomes frustrating, particularly when one pupil is asked to make a circuit from another pupil's records. They can be asked to suggest alternative symbolic methods. One such, a constructor's code, is incorporated in our programme. The desirability of an international convention for drawing circuits can then be presented and the conventional electrical symbols and circuit diagrams introduced.

* **Examining lamps**. Pupils dismantle lamps from the teacher's collection of defective lamps in order to trace the internal wiring. Fuses, made of filaments from an iron pot-scraping pad, are used to show the heating effect on which lamp filaments operate.

* **Bringing together what pupils now know**. In this stage, students reflect on the things they now know, or would like to know, about electric current. Material comes from class activities, from experiences at home, and from discussion with other pupils.

Table 10.1: Some suggested activities for the focus phase of the teaching sequence.

The pupils' own views: We have successfully used a whole-class discussion led by the teacher as the main strategy to make pupils aware of the different types of model that others hold for electric current (Chapter 2, Figure 2.16). Sometimes no one in the class suggests the scientists' viewpoint. When this happens the teacher can propose it, as an alternative, for example, to one suggested by another pupil in another class. Sometimes pupils suggest other models in addition to the four possibilities of Figure 2.16. Usually these turn out to be simply a variation of one of the four main models but others can be displayed if they do not seen to fit in with the four. Each pupil is asked to decide which model of those discussed he or she considers to be the best one.

Sharing ideas: One technique we have found effective, and therefore suggest here, is to pair together pupils with different viewpoints so that they can discuss and debate their preferred models. They can also assemble circuits and try to convince others of the validity of their views. Class discussion should follow and ideas can then be shared further. The numbers in the class preferring each model is of general interest to all class members, but the teacher should point out that the most popular view will not necessarily be the most valid one.

Designing the test situation: The teacher now leads the class to the need for an empirical test to decide the relative merits of the various models. An ammeter is introduced as an instrument that measures the size and direction of the electric current. It is necessary to demonstrate, and preferably to have pupils explore for themselves, how the reversal of the leads results in a reversal of the ammeter-indicated current.

Pupils and teachers need to discuss how the use of two ammeters (Figure 10.2) could be used to provide the critical test of the four models. We have found it useful to include one ammeter in the circuit first and for pupils to predict what the reading on the second ammeter will be according to *each* of the models in turn.

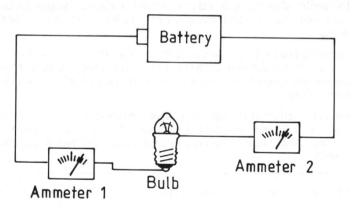

Figure 10.2: The critical two meter test. Centre zero meters are desirable but not essential. It is also possible to carry out the test with only one meter being used in the two positions sequentially.

The critical test: One view (Model D) survives the critical test. Unfortunately, however, this is the least intuitively acceptable model to most pupils. While, in our experience, most pupils will accept the evidence provided by the meters (albeit that they are 'black boxes'), many questions tend to be raised in pupils' minds: 'Why does the battery go flat after a time?', 'Why does the lamp glow if nothing is used up?' and so on. Other pupils want to reject the evidence or to use any slight difference in the readings on the meters to reinforce an alternative model. Most pupils are uncomfortable about the evidence provided by the test. They say it is just not sensible.

Helping the restructuring of ideas: Pupils therefore need the support of analogies and the introduction of other concepts, such as elementary ideas about electrical *energy*, to help them make sense of the evidence. We have tended to use a range of analogies; water flow (the current) produced by a pump (the battery) and flowing through pipes (the wires) and constrictions (the bulb); the pumping and heating of blood in the torso (the battery), the flow of this blood (electric current) to and from the extremities of the limbs where heat is lost (electrical energy); the flow of cars (the current) to the supermarket (the battery) to collect supplies (the electrical energy) and returning to the home (the bulb). The unfortunate but inevitable fact is that all analogies have some (often major) weaknesses, and this point needs to be emphasised. For example, a simple concrete analogy may lend an appearance of reality to energy as a substance (Warren, 1983). However, a more sophisticated or abstract analogy could result in more confusion than no analogy at all.

The application phase

The last phase of the teaching sequence is as crucial as the earlier ones if new ideas about electric current are to gain in plausibility and usefulness. We provide a set of practical problems at this point, which hopefully reinforce for pupils the advantages of holding a Model D (Figure 2.16) view of electric current and help them further clarify the distinctions between current and energy flow in simple electrical circuits. Pupils should be allowed to work at their own pace, usually in small groups, although some always prefer to work alone. Examples of the problems set out in our Teachers Guide are outlined in Table 10.2.

The success of this final stage is also dependent on active, intelligent teaching. Pupils need to be encouraged to try a variety of problems, and to extend the scope of those problems as they see fit. They often need help in clarifying the problem in their own minds, in seeing different ways it might be solved, and in reaching out for new ideas when they strike difficulties. More importantly, in terms of the central aim of the unit, they need help in seeing how a Model D view of current can be helpful in solving specific problems. Not only can such a view suggest a specific solution but it should be used to explain why it works.

Results from using the teaching sequence

We now turn to our experiences with the teaching sequence. These have involved using the sequence ourselves, as well as observing its use by other teachers. We have been invited into many classrooms while the electric current unit was being taught, and have interviewed a large number of pupils about their experiences in using the sequence as well as interviewing teachers about their pupils' performances.

*   **Solving the Christmas Tree light problem**. If one lamp in a line of three lamps (in series) is defective, none of the lamps will glow. The pupils are asked to find other ways of arranging the three lamps, so that if any one of them fails the others will still glow. The expected answer (a parallel arrangement) is usually obtained, but often only after considerable experimentation. Some novel solutions are proposed, and reported to the class.

*   **Simulating a set of traffic lights**. First, one set of lights is set up so as to be manually operated by a set of switches. Then a second set of lights has to be included in the circuit, without extra switches, to control a two way intersection. (Coloured filters can be used with the various lamp bulbs.)

*   **Designing a doctor's call-code**. A set of three lamps is provided to produce seven different on-off combinations, for a code of lights in a hospital foyer.

*   **Lighting a tunnel**. The bulbs and switches must be arranged so that a person walking through a tunnel can turn on a lamp for the region of the tunnel where he or she is, then, proceeding further a second lamp is turned on, and the first lamp is turned off.

*   **Simulating the lighting circuit for a car's lights**. Once complete, pupils are asked how indicator lights might be made to flash on and off.

*   **Caller indicator for the deaf**. The deaf person should be able to see, by looking at one, or two, bulbs, whether a visitor is at the front or the back door of the house.

*   **Operating an electric motor**. Using a length of resistance wire, make the motor go faster or slower. Design a method of reversing the direction of the motor's rotation, using switches.

Table 10.2: Some possible activities for the application phase of the teaching sequence.

Preliminary phase: Whenever possible, we began by using in-service education sessions to help teachers reconsider their ideas about the teaching of electric current. We have presented them with children's and scientists' views, and then helped them to clarify their own ideas. In these training sessions we have found that a video-tape showing someone actually teaching the challenge phase of the topic to be most useful. When such in-service education opportunities are not available adequate preparation using teachers' guide material only is somewhat more problematic. There is a very real conflict between attempting to provide sufficient background information and not overloading teachers with too much reading, as was mentioned in the last chapter.

Focus phase: Our experience has been that the main problem with many units of work on electrical circuits provided for young children is that pupils do not become sufficiently familiar with connecting up circuits. In our original interviews, we asked pupils to connect up a simple battery-bulb circuit. Frequently, as we saw in Chapter 2, they failed to do this successfully despite many of them telling us they had done it before in class. One of the main aims of the focus stage is to provide many opportunities for pupils to connect up electrical circuits (e.g. the Morse Code project). Subsequent interviews with pupils who had experienced the focus stage activities showed, even up to a year later, that almost all of them could quickly and efficiently connect up a simple battery-bulb circuit.

On the other hand, it is also clear from our studies that experience alone does not lead to a Model D view of electric current. Many 11 and 12-year-olds continue to hold a Model B view after the focus stage. Our interview with Peter, described in Chapter 2 and Figure 2.8, is an example of this. When he connected two bulbs in series he was again asked about current flow. This resulted in a complicated explanation, but one which was basically just an elaboration of Model B.

However, pupil-pupil interaction during the focus stage can sometimes lead to marked and sustained changes in children's views. Frequently pupils working in a group develop a common way of thinking about circuits. As a consequence a class of children at the end of the focus stage tend to hold one or two popular views, typically Model B and/or Model C. Table 10.3 shows this for two classes whose distribution of responses can be compared with a more representative sample.

11-year-olds	Percentage response for each model			
	A	B	C	D
Representative Sample (N = 302)	4	38	32	26
Class A (N = 29)	0	51	29	10
Class B (N = 15)	0	7	86	7

Table 10.3: Percentage responses immediately prior to the challenge stage compared with a representative sample of 11-year-olds.

Challenge phase: A non-quantifiable but very noticeable effect, from the viewpoint of both teachers and classroom observers, was the considerable interest all pupils showed in learning about the different ways others think about electric current. Whether they be 8-year-olds or university physics students, our experience has been that everyone is fascinated, and to varying degrees personally challenged, by a description of the four models.

However the problem is to help pupils change their idea toward the Model D view. We found that we initially underestimated the time needed to familiarise pupils with the ammeters, and to help them accept that the meters themselves did not perturb the system. Nevertheless, results from the two classes referred to above typify what happens after the discussion, debate and experiment (Table 10.4). They show that the majority of pupils now select Model D.

Finally, we have found it necessary, with respect to the challenge phase, to introduce the analogies suggested earlier so that pupils can understand why

Model D is plausible. Electric current has to be distinguished from, and contrasted with, the pupil's intuitive concept that something must be transferred from battery to bulb in one direction and then used up. We have attached the label *electrical energy* to this something, appreciating the danger in this and knowing that this abstract concept cannot be fully understood at this stage — if ever (see•Warren, 1983).

11-year-olds	Percentage response for each model			
	A	B	C	D
Class A (N = 25)	0	8	28	64
Class B (N = 15)	0	0	14	86

Table 10.4: Percentage response after the challenge phase.

Application phase: One of the biggest problems we have found has been in convincing teachers that this final phase of the teaching sequence is necessary to repeatedly reinforce, through pupil-teacher interaction, the circulatory-conserving view of electric current. In our experience whether or not there is a formal syllabus, or examination system, there is a strong drive by many teachers to proceed directly from completing one new topic or new concept to introducing another. Many teachers seem to find it uncomfortable to spend time consolidating a new idea, in spite of the demonstrable importance for subsequent learning of more complex ideas on that topic. During our Learning in Science Project research many pupils and ex-pupils have commented on this. For example:

You might be doing experiments in physics and you just start to get to know what you're talking about and they change it and do biology. . .then you forget everything that you know. . .in the end you don't know what you are doing.

ex-pupil

However, where pupils did become involved in an application phase, this was the aspect of the teaching sequence that they frequently thought was most significant for them. Stephen (aged 11) made the following comments in an interview he was not expecting, some 8 months after completing the teaching sequence.

Stephen   : *The best point was the puzzles. One of the best ones I think was the lighting system for the tunnel where you went through and turned on a light and you went a little bit further and if you pulled a switch the light behind went off and the one in front of you went on.*

Researcher: *Can you remember the circuit you made for that* (some 8 months previously)?

Stephen   : *Yes.*

Researcher: *Could you draw it?*

Stephen   : *Sure, it will be a bit messy.* (Figure 10.3 was drawn without hesitation.)

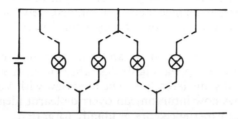

Figure 10.3: Stephen's solution to the tunnel problem.

Stephen   : *The thing is it does have its disadvantages. I mean, if one person went in the same entrance twice it wouldn't work. That's about the only problem with it.*

Justine (aged 14) is another example of a pupil who gained considerable satisfaction from the problem activities.

Justine   : *They were good activities in that you had to find your way through them yourself and find out which was the best solution... which was the quickest... the easiest... and I think you did end up finding your own answers... which I think is pretty good. It wasn't all sort of fed to you.*

Researcher: *I suppose one of the aims of teaching is to encourage people to learn how to think?*

Justine   : *Yeah... that's what this has really done for me... I found myself trying to work out different ways of doing it... by myself, you know.*

Stability of ideas over context and time: In the second edition of the Teachers Guide we have provided a set of written questions suitable for pupil assessment. Three of the ten questions were specifically designed to determine independently which model pupils would tend to use when they thought about electric current. The collective results from four classes of average to above average 14-year-olds (N = 115) showed that, on completion of the teaching sequence, 50% of pupils used a Model D view relatively consistently, 20% tended to use a Model D view in some situations but not in others, while 30% of pupils chose not to accept the Model D view or were unable to apply it to any of the situations discussed in the questions. These classes had in fact spent minimal time on the application phase of the work, and it can well be said that these results reinforce the necessity for the circulatory-conserving view of current to be discussed in a range of situations. It can also be argued that the results confirm the difficulties of changing children's ideas in a way that will be maintained.

The stability problem is further exemplified in Table 10.5, which illustrates what happened to the views of one group of 11-year-olds. It shows that, while the children appeared to change their views immediately following the lesson sequence, there was a certain amount of regression toward intuitive ideas over a period of time.

Where ideas are counter-intuitive and not reinforced by other learning situations, then it would appear highly desirable for further experiences with the topic to occur at some later date. The interview with Vanessa, discussed in Chapter 8, shows how intuition can override learnt ideas as some of the details of what was experienced are gradually forgotten.

With the reduction in the number of formal examinations that pupils face, there seems to be a tendency for teaching units and their associated assignments and tests to become self-contained sections of work. The results cited below,

| Class (N = 15) | Percentage response for each model | | | |
11-year-olds	A	B	C	D
Before lessons	0	7	86	7
After lessons	0	0	14	86
One year later	0	13	40	47

Table 10.5: Changes in the views of a group of 11-year-olds over time.

though not demonstrably representative, support the argument that pupils should from time to time be presented with problems which require them to recall and/or revise ideas taught earlier in the year.

Wider issues

We have discussed what happens to the learner when a specific unit of work on electric current, and the teaching sequence which underlies it, have been used in some classrooms. However, the effectiveness of the teachers' guide material cannot be assessed in isolation from the teacher who uses the material, since the teaching approach we have suggested requires active teaching. How teachers use any set of teaching suggestions will inevitably differ in some ways from what was intended by the writer of the materials. Sometimes these differences will be extreme, as can occur when the proposed teaching programme makes suggestions which go beyond the constraints of the hidden curriculum, and/or the motivation-control structures which operate in the classroom. For example:

(i)   Teachers are all busy people with only limited time for preparation. Some of us who have taught the topic 'electricity' for some years may not necessarily have a scientist's view of electric current, let alone a familiarity with children's views. It can be quite difficult to persuade some colleagues of the need to acquire a clear appreciation of the scientists' and children's views, and the contrasts between them. Further, to explore pupils' views with one's own class can all too easily be regarded as a waste of valuable teaching time.

(ii)   As teachers we can be unaware of the hidden curriculum which is operating in our own classrooms. For example, the tacit purpose of the classroom can be reduced to pure information transfer; verbal statements without real meaning. Pupils can easily recognise that regurgitation of this information is adequate as far as assessment is concerned. Where a thin fabric of teacher-control is embodied in this tacit process, any alternative teaching programme can potentially threaten the stability of the teaching-learning equilibrium. Both pupils and teacher can feel unsettled if the teacher stops telling pupils the answers, without giving them opportunities for adequate assimilation and elaboration!

(iii)   Teachers may also lack confidence and competence in certain topic areas. For biology and chemistry graduates, as well as non-science graduates, electricity topics can often be a good example of this. In such situations the more open-ended work of the exploratory and application phases can build up to a threatening situation and once more distortion of the proposed teaching programme can result.

It is problems such as these which have been summarised so effectively by Erickson (1982, p. 172) who states:

. . . curriculum displacement does not reduce to a series of logical implementation procedures once 'the product' is delivered by the developers. It is rather a very complex process of decision-making that is continually mediated by inputs from broad social factors in combination with much more situation specific factors which reflect the realities at the local district or classroom level.

We have chosen to emphasise these issues because they are central and basic ones for any curriculum change. Should one first attempt, for instance, to change teacher philosophy in a general way and allow teachers to work through to implications for their own programmes? Or should teachers be encouraged to try out a unit which incorporates what is for them a new philosophy? In the first case the lack of specific exemplars can lead to misunderstanding, while the second may be doomed to failure without a change of approach on the part of the teacher, or simply because the specific unit does not match the teacher's strengths and interests in some otherwise superficial way. While the answer may be to proceed along both lines simultaneously, the limited time available for teacher in-service education ensures that the task is not a simple one.

Part 5

Implications for Curriculum and Teacher Education

11 Implications across the Curriculum

Peter Freyberg

Since we have been, up to this point, primarily concerned with the learning and teaching of science, most of our discussion and illustrative material has been related either to the general science curriculum or to topics in physics, chemistry and biology. In this chapter we would like to demonstrate, from research in other curriculum areas which has developed out of the Learning in Science Project, how similar pedagogical problems arise in other subjects. We will consider these problems under three headings — children's meanings for words, communicational mismatch and 'mini-theories' — and will give examples mainly from the fields of mathematics, social studies and earth science.

Children's meanings for words

We have already given numerous examples (particularly in Chapter 3) of the ways in which pupils can misconstrue their teacher's intention because they interpret a word which is being used with a restrictive, specialised meaning (for example, insect) in a more general everyday sense — or vice versa. This does not happen solely with science topics, of course, although it is somewhat more likely to occur there because of the scientist's need for unequivocal terminology.

Take the word 'community', for instance. This is frequently used, in its everyday sense, in social studies lessons: it is in fact the label for one of the major recurring themes in that subject. In ecological studies, however, 'community' is used in a much more restrictive way to describe two or more different species coexisting in a particular relationship to one another. In New Zealand schools the curriculum is likely to include community as a topic in the same year in both social studies and science. Is it surprising, then, that children sometimes become confused as to which meaning is intended — the general or the restrictive? Major problems occur when particular lessons or sequences of lessons proceed with the teacher assuming that, whenever she uses a particular term in that type of situation, all her pupils will interpret it in the same way that she does. How can a child who thinks that a community can consist of any group of animals living together — a herd of cows, for instance — really grasp the concept of ecological balance?

We have observed even more far-reaching consequences of students failing to learn the difference between restrictive and general uses of a word — in mathematics we have come across teacher trainees whose concept of a fraction was limited to vulgar fractions such as ¼ or $^{47}/_{79}$. They did not consider a decimal such as .25 to be another type of fraction. Indeed they did not conceive of decimals as representing 'parts of' anything. In spite of the fact that they recognised the equivalence of ¼ and .25 at one moment, at the next they maintained that decimals and fractions were entirely different from one another. No wonder that even their computational competence in this area was affected.

Similar possibilities for confusion exist in the initial stages of teaching in the earth sciences, as Happs (1982a, 1984) discovered while working in our Science Education Research Unit. He was concerned with children's ability to observe and classify a selection of rock and mineral samples, but very quickly found out that their preconceptions of what was a rock were an important determinant of their actions.

To an earth scientist, *rock* is a material which consists of minerals, whilst *a rock* is taken for granted (by the earth scientist) as being a fragment of rock,. irrespective of size, shape, colour or density. Happs interviewed 34 11 to 17-year-olds : those up to 14 years of age were taking both social studies and science, while the older pupils were studying geography and at least one of physics, chemistry or biology.

Most of the students Happs interviewed held very different views from the earth scientist. Their classification criteria were based on just those which scientists would generally consider to be irrelevant — weight, colour, size and jaggedness. For example, a fist-sized sample of dull-looking, jagged greywacke was readily described as 'rock' by 80% of the interviewees, but only 28% acknowledged a small rounded piece of pumice as 'rock'. To many students the pumice was 'too light to be a rock'.

What I call a rock — I go on weight. I don't think it is right scientifically, but I just go on weight.

17-year-old

A third of the students who were interviewed labelled small rock fragments as stones which they distinguished as being different from rocks, although 'stone' is a word which has no widely accepted meaning as far as scientists are concerned.

In a subsequent investigation Happs (1984) endeavoured to tackle this problem of different meanings head-on, by providing pupils with a simplified definition at the outset. His definition included the terms 'solid, natural substance'. Unfortunately for some children this simply reinforced their own categorisation systems. Pumice was not previously considered by some pupils to be rock because it was light: now they could justify their earlier classification by saying that it was not really solid. Likewise a polished piece of granite, which was previously considered not a rock because it was too smooth, gained the additional justification (in the pupils' eyes) of being 'not natural' i.e. it was man-made in the sense that it had been fashioned by humans. Fowler (1966) has discussed similar difficulties inherent in all definition-oriented curricular materials, through pupils construing a definition to fit their intuitive ideas about instances and non-instances.

Even in mathematics, where we would have thought that the context would limit possibilities of misinterpretation, we have found some striking discrepancies in the meaning attributed to words by the teacher or textbook author, and by at least some of the pupils (Carr, 1982). For instance, the question 'What is the sum of 8 and 6?' was answered variously as 2, 14 and 48 by different children. Careful questioning revealed that the wrong answers were occasioned not so much through confusion as to *which* arithmetical process was involved as through interpreting 'sum' in the general sense ('doing your sums'), which would allow *any* of the operations to be carried out.

Similarly, Otterburn and Nicholson (1976) discovered that a substantial number of CSE (16-year-old) pupils in the United Kingdom interpreted 'product', used mathematically in its everyday sense, as something that has been produced. Perhaps this could account for their finding that 20% of their sample of 300 children interpreted 'product' as meaning 'sum' or 'difference', which would parallel our New Zealand naturalistic research.

Elsewhere (Freyberg and Osborne, 1981) we have drawn attention to other ways in which children's different meanings for words can have pervasive effects on their subsequent capacity to take out of a lesson what a teacher intends. For instance:

(i)   Children may construct a meaning for an unfamiliar word simply on the basis of verbal association or verbal similiarity. One child we interviewed about measurement believed that a kilogramme was a measure of distance, like kilometre, because they both had the same initial morpheme.

(ii) Even those children who appreciate the differences between a technical (restricted) meaning of a word and its more general meaning can fail to recognise that some terms have two or more different *technical* meanings, each appropriate only to particular contexts. We have found pupils who have associated the conservation of energy talked about in social studies with that studied in physics, with inevitably confusing results since the former is applicable to open systems and the latter only to closed systems.

(iii) Children may construct ephemeral meanings for words with which they are unfamiliar, based simply on adjacent context 'clues'. In one lesson we observed, a class of 14-year-olds were plotting various physical features, including ocean trenches, on their individual copies of a map of the Pacific Basin. One child had marked an ocean trench continuously alongside the major land masses because it was the 'boundaries of the ocean'.

In each of the examples cited here — and we could give many others — we have been able to identify the child's underlying conception because we were talking to *individual* children. Unfortunately, with class sizes as they are, this is a luxury beyond the reach of most teachers. Some misinterpretations of words (from the teacher's point of view, of course) are picked up by teachers in class questioning; others become obvious in the course of subsequent lessons. We suspect, however, that a surprising number persist for months or even years, as evidenced by the failure to recognise decimals as a kind of fraction which we mentioned earlier.

Communicational mismatch

There are other reasons why pupils sometimes misunderstand what their teachers are trying to tell them, apart from those just discussed in relation to concept labels. We all use words occasionally which are outside the vocabulary range of our audience, for example. Some teachers do this intentionally as a means of extending their pupils' vocabulary: this can be a useful strategy provided that the unknown/known information ratio is kept low and children are encouraged to ask about any words with which they are unfamiliar. The real problem occurs when teachers do not think that there is any possibility of a misunderstanding, because the language being used is all familiar to children of that particular age-group.

Nevertheless, as Barnes (1976) has pointed out, children can fail to pick up what the teacher intends because they are operating on a different set of discourse rules. For instance, what the teacher intends as a question children can take as a statement of fact. Where the question anticipated a negative answer, the children will learn the opposite of what was intended.

Similarly with metaphors, which are scattered liberally through most discourse. We talk about 'jumping the gun' in making up one's mind before all the evidence is in, about 'seizing' an opportunity while it is there, and about 'burying' an idea which is not useful. Adults have learnt to cope with the problems of interpretation where multiple meanings are concerned, sometimes by deliberately suspending judgement until subsequent discourse points to one or other meaning as the most probable but more often by utilising subtle linguistic or other contextual clues. Pre-adolescent children, however, usually have difficulty in recognising when a metaphorical rather than a literal meaning is intended. Analogies, too, can lead to unanticipated 'meanings' being acquired — not only verbal analogies but those implied in diagrammatic representations. For instance, in the course of our preliminary investigation of children's understanding of decimal fractions, we came across several pupils in different schools who thought that .49 could, in some ways, be considered larger than .94.

When we pursued this puzzle further we found that these children appeared to be operating at the same time within two parallel conceptual frameworks, illustrated by the number lines in Figure 11.1.

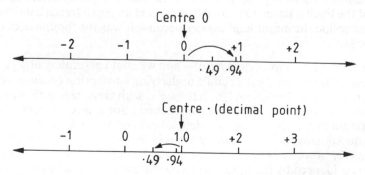

Figure 11.1: Problems with number lines (after Freyberg and Osborne, 1981).

On the first line these children see .94 as larger because it is further away from the centre: on the second line they see .49 as larger for the same reason, the centre this time being the decimal point. To explain how this could happen one has only to look at the possible confusion of diagrams occuring at their different places in their textbook (Figure 11.2).

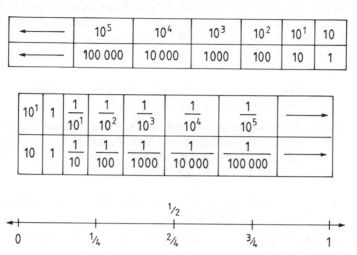

Figure 11.2: Textbook diagrams.

In this case, children have carried over one framework provided by the textbook to an inappropriate situation. In fact, the arrow which is commonly used in school textbooks in many subjects would appear to be a fruitful source of pupil misconception, as Schollum (1983) has discovered.

The overall context within which the teaching communication takes place can also have far-reaching effects on what is learnt. Here we are referring not to the immediate context surrounding a specific lesson but rather to the continuing context of the classroom — the day-to-day exchanges usually undirected towards any particular teaching goal. Some examples from research in the teaching of social studies will illustrate this point.

In his exploratory work in the field of social studies, Smythe (1983) identified numerous parallels with the findings in the science area. He used a modified interview-about-instances or interview-about-events approach to investigate children's understanding of some of the concept-labels commonly used in social studies teaching e.g. 'co-operation', 'conflict', 'customs', 'traditions', 'social change' and 'values'. In the first section of the interview, children were shown photographs which they were asked to sort into examples or non-examples of each concept, explaining why they did so. In the second part they were asked to give instances of the concept from their own experience or from their reading, TV watching and so on.

Two particular aspects of Smythe's research are of interest here: first, the pervasiveness of the we-they categorisation which children develop and its influence on children's interpretation of social science materials, and second, the values associated with concepts such as co-operation, conflict and change, from early childhood onwards. Once again the observations are by no means novel — a great deal has been written on children's egocentrism, for example — but what Smythe's comprehensive interviews with 97 school pupils aged 7 to 14 years bring out is the extend to which their prior ideas would most likely affect their interpretation of their teachers' comments in social studies lessons.

Smythe found, for instance, that references to traditions and customs were usually interpreted, by the younger children in particular, as drawing attention to the behaviour and celebratory or ceremonial practices of *other* people — not themselves and their own families.

Interviewer: *Well, Garth. . . where did you get your understanding of customs and traditions from?*

Non-Maori   *Well, we usually do lots of things on Maoris* (the pre-
child   : European inhabitants of New Zealand). . . *so we learnt about customs and traditions last year. . . that's how I kept on getting to learn these.*

Many of the children also showed that they associated customs and traditions with the past rather than the present.

I suppose a custom and tradition is to keep it exactly the way it was. . .from olden times. . . Things like the opening of Parliament, they always do that every year. . .just the way they do it from when Parliament was first begun. . .

14-year-old girl

Similarly, when asked why it is that we have traditions, an 11-year-old girl gave a typical response:

Child   : *To. . . well, kind of honour their ancestors and that kind of thing. . . try to carry on what they had. . .*

Interviewer: *Why do they want to do that?*

Child   : *Because they feel it's a good thing to do. . .they feel. . . well, their ancestors would be pleased.*

As Smythe points out, the problem facing social studies teachers is, first, how to broaden children's concepts of culture, custom and so on, so that these include their own modes of behaviour as well as those of others; and, second, how to use examples of cultural diversity to demonstrate cultural similarity — the same needs satisfied in different ways. However, it is unfortunately possible for children to maintain their notions of difference and implicit superiority of their own culture at the same time as they talk about cultural similarities and equalities of human rights. If it takes more than verbal teaching to convince children that some of their intuitive *scientific* ideas will not adequately explain what is going on about them, how much greater will be the requirements for insightful teaching which takes into account children's existing ideas in the more emotionally-charged topics of social studies.

A somewhat different theme also investigated by Smythe was children's understanding of the terms 'co-operation' and 'conflict'. The first of these is part of the vocabulary of most classrooms, from the beginning of formal schooling upwards. 'Conflict', Smythe found, was not a commonly recognised term until the age of about 13 or 14 years, but even the youngest children recognised the opposite of co-operation although they could not put a label to it.

The most significant aspect of the research, from our point of view, is the strong positive value children accorded to the notion of co-operation, from the earliest school years, and the strong negative value associated with its opposite. Co-operation was assumed to be a 'good thing' regardless of the activity concerned, and non-co-operation was always assumed to be 'bad'. And the most common example which children gave of co-operation was 'doing as you're told'.

We have already discussed in some detail that bug-bear of science lessons, the pupil's need to 'get the right answer', regardless of whether that answer corresponds with the individual's observations or not. We see the research on co-operation as pointing up a similar danger. Just as we want the young scientist to rely on his observations as well as on his expectations — and to attempt a reconciliation if they disagree — so we would want children to look critically at situations calling for co-operation, to see whether or not that co-operation is justified on other grounds, for example, the worthwhileness of the activity.

Similarly, we would want children to question their notion that conflict is always wrong, and that it necessarily involves physical or verbal aggression. Out of the conflict of ideas new knowledge often emerges, out of disagreement a compromise course of action. If we are to get across to children a rational basis for understanding their social as well as their physical world, we need to provide them with *real* opportunities to test their ideas, through open expressions of opinion subject to the same kind of scrutiny as we would hope them to give the observations they make in science lessons.

Mini-theories

Most human behaviour, even that of very young children, is surprisingly self-consistent considering the welter of different stimuli which the person experiences. Our actions are based on a continual flow of predictions which we make as to what is going to happen next. The whole process is largely subconscious, particularly in adults, but occasionally surfaces when we stop and ask ourselves why such-and-such occured in that particular way.

Children as young as 2-3 years of age are reflective in this way, as is illustrated by the frequency of their 'why' questions. This seeking for explanations represents the conscious tip of the predictive iceberg — our attempt to impose some sort of order on our environment. In other words, we learn to act *as if* the world had certain characteristics related to one another in a specific manner: we learn to act in accordance with a complex series of hypotheses — 'mini-theories' as Claxton (1984) calls them, in relation to young children.

In many ways every concept that we hold incorporates a mini-theory: its complex structure, its particular exemplars, the connotations which its label (word) evokes, all contribute to the picture of how we perceive that specific portion of our world. Thus it is sometimes possible to deduce, from the meanings which children give for particular words, something of what lies behind those meanings — their overall cognitive structures. In our LISP research this has enabled us to extrapolate from children's stated meanings of words like 'force', 'animal', 'burning' to their perceptions of underlying processes or contributing factors, to their classification systems and so on.

Again the observations we have made in the various science teaching areas — biology, physics, chemistry — can be paralleled in other curriculum fields. Probably the clearest examples of how pupils' mini-theories subvert the learning which their teachers intend are to be found in mathematics.

Our colleague Ken Carr provides some typical instances. He reports (in Osborne, Smythe, Biddulph and Carr, 1982) that a third of the 11 and 12-year-old children he interviewed believed that decimal fractions mirror the place value nomenclature of whole numbers. He quotes:

Interviewer: *What would we call the next column to the right?*
You've got hundreds, tens, ones, the decimal point.
What would the next column be?

Pupil : *Ones. You just go in reverse again. Ones, tens, hundreds and just carry on. (Why?) You've just turned it around.*

Interviewer: *So what does the decimal point do?*

Pupil : *Reverses the numbers. From then on you reverse the number.*

Interviewer: *Why is our number system like that?*

Pupil : *I'm not sure really. It's just making it easier to count with. So it's understandable to most people. So we can understand it . . .*

In a later investigation Carr (1983) found only 8 out of 28 pupils aged between 12 and 13 years who could correctly write the decimal fractions for three tenths, seven hundredths, fifteen hundredths, seventeen tenths and twenty hundredths. He cites one interview:

Interviewer: *Seven hundredths. Could you write seven hundredths* (presents to the pupil a card with seven hundredths written on it).

Pupil : *Point oh oh seven* (writes '.007').

Interviewer: *How did you work that out?*

Pupil : *Well, I've done it in hundredths. Hundredths is two zeros. So I put decimal, two zeros and then the seven.*

Interviewer: *Why?*

Pupil   : *Well, a hundred has two zeros.*

This appears to be by no means a unique or culturally-bound situation. For instance Brown (1981) reports that in her United Kingdom study 50 percent of 12-year-olds and 40 percent of 13-year-olds could not write the decimal fraction 'three hundredths'.

In addition, the work of Erlwanger (1975), Ginsberg (1975), Davis (1979), Carpenter and Moser (1979), Lawler (1981), Hart (1981), Moser and Carpenter (1982) and others has identified a variety of mini-theories held by children about mathematical phenomena. Carpenter and Moser (1979), for example, have found that many children learn to count and can successfully solve basic addition and subtraction word problems prior to formal instruction. In their view the strategies children employ result from their 'intuitions', based on action-experiences together with some informal instruction.

Elsewhere, Lawler (1981) has shown quite dramatically how children can have different mathematical theories for different situations. He describes different 'theories' his own daughter used for simple addition; each theory (or micro-world of the mind as he calls it) applied to a different set of situations. The procedures for adding numbers in the context of money, for example, were quite different from those used for adding numbers set down in columns of digits.

In mathematics, as in other subjects, children's intuitive ideas may indeed fit comfortably with formal teaching or they may not. For example, the notion that the multiplication of whole numbers is a shorthand for repeated addition develops readily out of their daily experiences for many children. Katterns and Carr (1984) illustrate this with a 7-year-old who considered that multiplication, or times, 'tells you how many things there are and gives you an answer a bit quicker'.

Child   : *Like four lots of five equals twenty, or four fives is twenty.*

Interviewer: *So what is another word for times then?*

Child*   : *Lots*

Interviewer: *Could you show me four lots of five?*

Child   : (Draws) *And it all equals twenty.*

However, other arithmetical procedures can be counter-intuitive. Consider: $\frac{10}{30} + \frac{40}{70}$ To many people, children and adults alike, it is obvious that this equals $\frac{50}{100}$ and indeed there are situations for which this result is correct. For example, if an examination consists of two parts, with a possible total of 30 marks for Part A and 70 marks for Part B, then scores of 10 marks and 40 marks result in a student gaining 50 marks out of 100 in all. Erlwanger (1975) and Knight (1983) describe both child and mature adult cases in which this 'rule' for the addition of fractions had been developed and retained for *all* such algorithms, in spite of formal mathematics teaching. We consider these to be significant instances of the way in which intuitive mathematical ideas, often developed at an early age, appear to be supported later by situations in everyday life.

Brown and VanLehn (1982) suggest that, when pupils find that these intuitively-based ideas do not work and reach an impasse (for example if $\frac{10}{30} + \frac{40}{70} = \frac{50}{100}$ is marked incorrect), they seek to rectify the situation with the most parsimonious modifications to their existing ideas. The result is a hodge-podge which still does not work. Similar erroneous procedures, or 'bugs' can also arise when pupils partly forget taught procedures and repair their deficiency by patching up what has been forgotten.

Finally, we believe that many, if not most, children hold a pervasive mini-theory which is probably at the root of their alternative conceptions in many curriculum areas i.e. that the world was always as it is now, and such changes as have occurred must have been sudden and comprehensive. *Gradual change* is a sophisticated concept, and one of the most difficult for children to develop.

In this connection, Happs (1984) has explored children's ideas about how glaciers, mountains and soil originate. For one investigation Happs (1982a), used a pupil sample similar to that used for exploring children's meanings for the word *rock*. He found that the majority of pupils believed that the soils in a local area were either formed when the earth was formed:

It has always been there. . . when the earth got created.

11-year-old

or originated from volcanic activity:

When you have a volcano erupting you've got lava coming out. . . it dries up and turns into stone. It could be something like that — comes up and turns into soil.

14-year-old

Volcanoes are familiar to New Zealand children, and the idea of soil originating from volcanic activity was also proposed by some of the older students. For example:

I think it (soil) would be uplifted. . . from volcanic activity.

17-year-old

Only six of the 30 students interviewed considered that soil is something that evolves on the site where it is found, and that it is the product of a continuing process rather than a single event. Similarly, most students interviewed about glaciers and mountains answered in terms of single acts of creation (for example, earthquakes) rather than in evolutionary terms.

Implications for teaching
Our research on children's concepts and modes of thinking in other curriculum areas has been neither as intensive nor as extensive as that in science.

Consequently we need to be more cautious about generalising from the results, and to regard explanations of what influenced those results as far more speculative.

Nevertheless, we consider that the research findings just cited, together with those from an increasing body of qualitative research being undertaken in many other countries, provide teachers and curriculum developers in all subject areas with a considerable challenge. The problem is common to all areas: how can we ascertain what our pupils already know, so that we can devise teaching materials and strategies which take this into account?

In the field of science teaching we have argued that there are some common preconceptions which pupils are likely to bring to any given lesson, and that the teacher should be aware of these while on the lookout for any idiosyncratic views. There is no reason to believe that the situation will be different in any other field. Since teachers cannot be expected to extract the research findings on children's ideas for every topic they will teach, this places a great responsibility on the curriculum developers and textbooks writers to ensure that this information is provided to teachers in a readily-available form. We ourselves are experimenting with appropriate materials in this respect.

What teachers in all subject areas can do, however, is to encourage a genuine and continuing interchange of ideas in the classroom, between pupil and pupil as well as between pupil and teacher. In practice it is only through such discourse that teachers can monitor what they are doing — written tests have their uses but they can rarely tell us *why* a child thinks this way, only that he *does* so. As Cockroft (1982) argues in the case of mathematics teaching, we need to provide more opportunities for pupils to talk about what they are doing, to become aware of their own ideas and those of their peers, and to modify their own ideas where necessary. This is the closest that we are ever likely to get to a teaching situation in which both teacher and learner expectations are fully realised.

12 Introducing Children's Ideas to Teachers

Roger Osborne and Ross Tasker

When we have talked to fellow teachers and teacher educators about the concerns described in the earlier chapters of this book, many have recognised immediately what we were referring to. These colleagues were already aware of the problems or could relate them easily to their own experiences and intuition about the learning of science, and of other subjects, in the classroom setting. However, others have initially found it difficult to accept that their assumptions about what children interpret from their well-prepared lessons could be so different from what they (as teachers) intended. Nevertheless, they too, after using our interview schedules and/or surveys with their own pupils, usually appreciate the false assumptions that we all have been making so often.

When teachers become more aware of children's ideas and the consequential difficulties pupils can have in learning science, they experience conflicting feelings as to what they can do about it. Some consider that a panacea may be readily available in some alternative teaching mode — for example, in individualised learning or contract work. Our observations lead us to believe that these methods provide no such easy solution; the essential interaction of minds — pupil to pupil and teacher to pupil — can be all too readily lost in a paper war of mastery tests and contract marking.

Another reaction is to shift the blame for the problem to some other sector of the education system. The problem, it is claimed, is the result of factors beyond the control of the teacher — external examinations, syllabus constraints, school organisation, class size, textbook constraints or lack of preparation time. Alternatively, it is commented that the problem would not occcur if pupils had been properly taught at primary (elementary) school, or if teacher-trainees had been taught effectively at teachers' college. Anyone involved in education can put forward this type of argument, but it does not get us anywhere.

A third type of reaction leaves some teachers simply despondent. They may feel that they have tried so hard, even to the limit of their abilities and energy, and yet it now appears that much of what they have done has been ineffective. Most of these teachers soon overcome their despondency and become interested in finding ways to make their teaching more effective.

Fortunately the most common reaction we have found is a positive one. These teachers are aware of the problems, at least to some extent, and are keen to do something about them. Typically, however, these have been already-committed science teachers. Where the teachers we talk to are not teachers of science there are other concerns. Questions are raised about the aims of science education, and about whether or not we need to, or should even attempt to, change children's ideas.

In this chapter we will discuss some of the workshop techniques used in the dissemination stages of the Learning in Science Project by ourselves, by science advisers, and by practising science teachers. The audiences have been teacher-trainees, science teachers and, on some occasions, teachers of subjects other than science.

Confronting teachers with children's ideas

We have found it very useful for intending workshop participants to make a prior survey of the pupils in their own classes on some simple and basic ideas. A typical survey is given in Appendix D1. Teachers are asked to bring their results to the workshop, where the data can be pooled, and graphs and tables of the collective results can be displayed.

An alternative procedure is for each participant to interview a child at the beginning of the workshop. A brief introduction to the technique of interviewing is provided (15 minutes), a set of instructions (Appendix D2) and interview diagrams are given out, and each workshop participant is introduced to a pupil from a class brought in from a local school. At the end of the interviews (typically 20 minutes) the participants pool their results and discuss the comments that pupils have made in the interviews.

Most frequently we use the same interview schedules for the workshop sessions which we have used in our major studies, since we then have survey data available which indicates the prevalence of particular viewpoints over a large representative sample. It is usually opportune to display and discuss these in the workshop at this stage.

Confronting teachers with their own ideas and the ideas of others

To introduce teachers to children's and scientists' ideas on a particular topic, and to help them clarify their own ideas, we normally use interview-about-instances cards in workshops. Our technique is to pair up participants, give them a set of interview-about-instances diagrams and ask them to come to an agreed position about each situation depicted, together with the reasons for that response. The pairs can also be asked to discuss together the likely response of pupils in their own classes to the questions on the diagrams. Sometimes we will ask the workshop participants to attempt one of our surveys themselves.

When this has been completed, we proceed to discuss the teachers' own ideas or, since it is potentially less threatening, the question of 'how I think my pupils would respond'. Usually this promotes a free ranging discussion. We then illustrate the ideas that pupils we have interviewed hold about the specific situations depicted on the cards. Where appropriate we also provide the results which we have obtained with the large representative samples when we have used similar survey questions. In the final phase of the workshop we discuss our experimental approach to modifying pupils' views with respect to the topic where, and if, we have such research data.

An alternative approach, when less time is available, is simply to give workshop participants the opportunity to complete the appropriate survey. Where ideas about pupils' views are to be discussed with non-science teachers a general survey has been used (Appendix D1). Results obtained from the workshop participants can then be pooled and compared to the previously-acquired results from a representative sample of pupils.

An example of how this procedure works comes from our experience with one of the secondary schools in our district (Tasker, 1982b). Senior science staff requested that a workshop on children's ideas should be given to the *whole* staff of the school and, after preliminary discussions with the school administration and the teachers concerned, this was agreed to. In the first activity of a one-day workshop, all teachers completed the survey illustrated in Appendix D1. Their results were immediately collated by senior pupils of the school and then displayed alongside the results from pupils at various class levels in the same school (Figure 12.1). These data provoked considerable discussion and debate, and the workshop leaders had to ensure that the central issues, as they saw them, were not lost sight of. These issues were:

- Both pupils and teachers hold a range of different ideas about objects and phenomena which are basic to science learning situations in classrooms.
- The ideas of specialists in a particular field are frequently different from everyday ideas, and teachers pay insufficient attention to this fact in their teaching.

Finally, where there is even less time available — in a one hour talk, for instance — we have provided each member of the audience with a copy of Figure 12.2. They are asked to think about their meaning for 'an animal is a living thing', and then to consider the list provided, placing a tick alongside those things that they consider to be animals, and a second tick alongside those they consider to be living. Each person is then asked to total the number of ticks placed in each column. A show of hands invariably indicates a wide range, particularly for the animal column. The point is made that we all construct our own meanings for, and elaborations about, words that we see and hear.

Activities with teachers

In some workshops, if time is available, we have deliberately exposed teachers to the problems inherent in many teacher-guided activity-based lessons by requiring them to actually carry out an activity in small groups. Where the workshop involves teachers from a number of different subject areas, not just science teachers, groups are probably best arranged to include a range of subject expertise, teaching experience and so on. The instructional material provided and the level of additional oral instruction is kept as similar as possible to the typical classroom situations we have observed.

As an example, Figure 12.3 illustrates the kind of instructional material we have used, which is a slightly modified version of an experiment found in a well-used New Zealand textbook (Petchell, 1976). The activity is designed for use with 13 and 14-year-old pupils. Workshop participants are introduced to the material in a succinct and explicit manner: 'You are going to carry out an

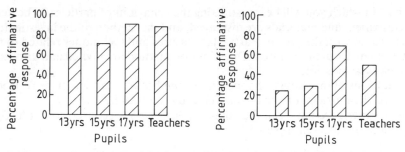

a Question: Is a person an animal?   **b** Question: Is a spider an animal?

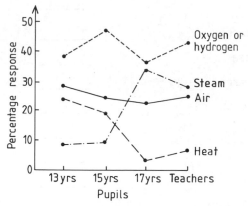

c Question: What are the bubbles made of?

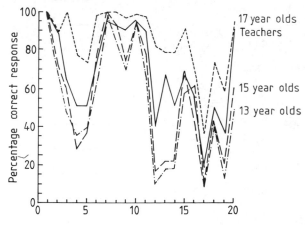

d Proportion correct on all survey questions.

Figure 12.1: Pupils' and the teachers' responses to the survey questions of Appendix D1 obtained in one New Zealand secondary school. (Note: The 17-year-old pupils included students who were studying at least one science subject.)

activity in which you will be investigating the behaviour of particles.' The items of equipment and materials are displayed, and the method of the investigation is discussed with the group. Particular features and techniques related to the experiment are demonstrated. Questions on method are invited and answered as clearly as possible.

Questions about what is happening in the beakers can either be reflected back to the participants or be answered by the 'teacher'.

Our experience has been that these alternative procedures do not lead to significantly different results.

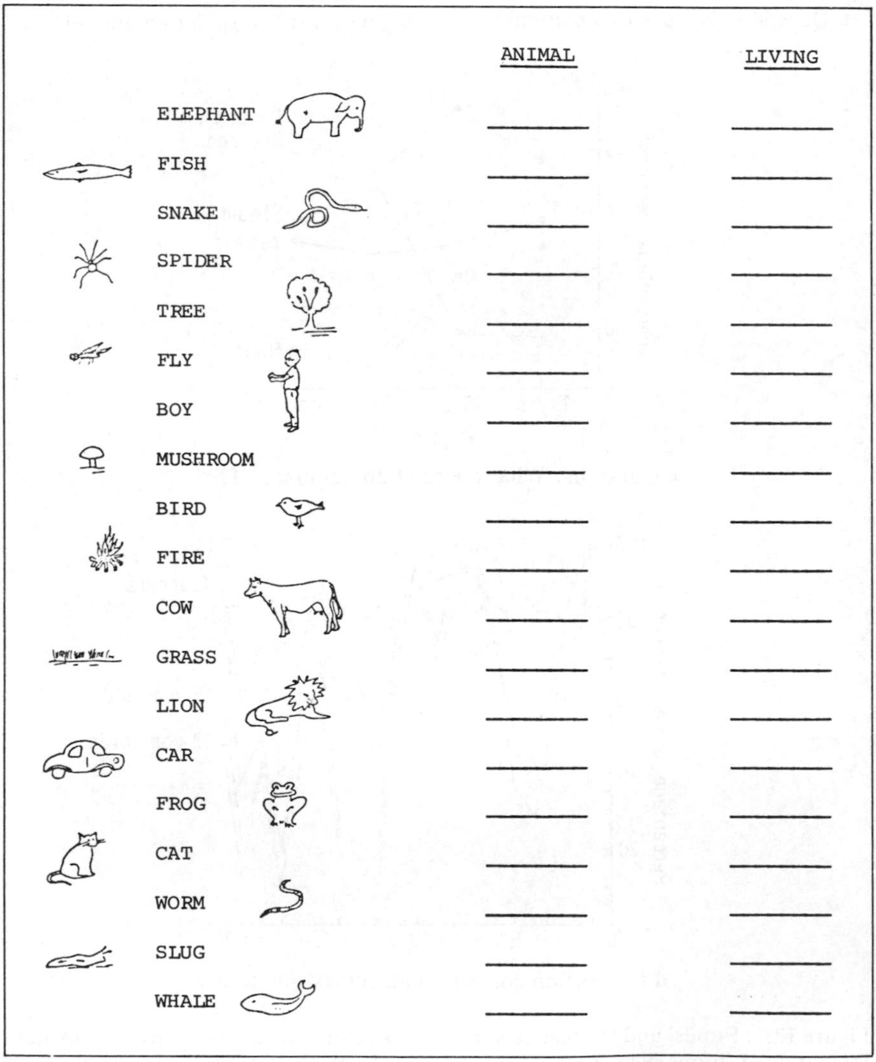

	ANIMAL	LIVING
ELEPHANT	_____	_____
FISH	_____	_____
SNAKE	_____	_____
SPIDER	_____	_____
TREE	_____	_____
FLY	_____	_____
BOY	_____	_____
MUSHROOM	_____	_____
BIRD	_____	_____
FIRE	_____	_____
COW	_____	_____
GRASS	_____	_____
LION	_____	_____
CAR	_____	_____
FROG	_____	_____
CAT	_____	_____
WORM	_____	_____
SLUG	_____	_____
WHALE	_____	_____

Figure 12.2: Exploring teachers' meanings for 'animal' and 'living'.

When members of the group have finished the investigation they complete a survey (Appendix D3) designed to expose ideas held rather than simply to identify whether or not the participants' thinking matches the scientific viewpoint. The results can be collated and compared with the results obtained from a group of 13 or 14-year-old pupils in the school (Figure 12.4).

Investigating the behaviour of particles

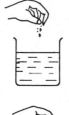

You have already seen some of the things that the particles making up crystals can do when they move from solution back to the solid state again.

In this investigation, if you use some crystals whose particles are strongly-coloured, you will be able to observe how the particles move when they go back into solution.

Equipment needed: beaker of hot water, beaker of cold water, Condy's crystals.

Make sure the water has settled and is not moving at all. Drop one large crystal into each beaker. Watch carefully. What happens to the Condy's in each beaker to start with?
What happens to the Condy's in each beaker after a while?
What was the most important difference between the beakers to start with?
What do you think was the most likely cause of the different results?

In your notebook:

(a) Write the heading.
(b) Draw a diagram of the two beakers. Carefully shade in the Condy's to show how it looked at the end of five minutes.

Copy and complete these sentences:

(c) I think particles in water (can, cannot) move about.
(b) These particles move (more easily, not so easily) if the water is hot.

Questions to discuss:

Particles, like other objects, cannot move unless they are supplied with energy. How can the different amount of movement in the hot and cold beakers be explained?

Figure 12.3: A typical science activity worksheet (after Petchell, 1976).

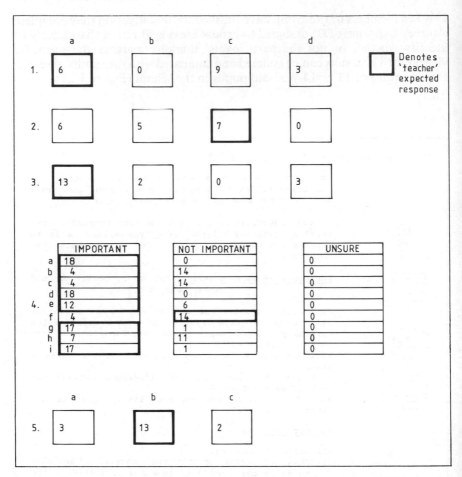

Figure 12.4: Typical responses to the survey questions of Appendix D3 (N = 18).

In our experience teachers placed in the position of learners react in the same way to carrying out an activity as do pupils: they invent purposes not intended, they copy the actions of other workshop participants in whom they have confidence, they attempt to find what answer it is that the teacher would like them to get, and in the process of obtaining answers they create the same kinds of mess with equipment as do a group of 13-year-olds.

In the resulting discussion it should be possible to bring out a number of points bearing on the difficulties of learning in classroom situations, as discussed in Chapter 6. For instance:

1   Lessons are perceived by many learners as isolated events. They do not consider specific aspects of the lesson against relevant prior experiences.
2   The lesson's purpose is not understood by many learners. They invent what seems to them to be a reasonable purpose.

3 Design features of a lesson are not appreciated by many learners; they are quite unaware of which aspects of the lesson need particular attention and why.
4 Teachers make wrong assumptions about the views many learners hold.
5 Learners frequently do not see the significance of their lessons in the way teachers think they do.

Discussion groups

The issues we have raised throughout this book provide a wealth of topics for teachers to debate at conferences, at workshops, and in staff meetings. Through such debate we begin to appreciate the other points of view that individuals hold, and to understand better our own views in this context. Just as it is difficult for pupils to change their ideas about the meanings for words, and about how and why things behave as they do, so it is difficult for teachers to change *their* views on teaching and to modify the instructional strategies they have developed for themselves. What we have attempted to say in this book about science teaching can and will be misinterpreted sometimes, just as the ideas we all attempt to teach in science are often misinterpreted. It is only through discussion with others that such misinterpretations and reservations are realised for what they are.

The threatening effects of confronting one's own possible misconceptions should not be minimised. Some responses from teachers which followed an introduction to our work given by their own Head of the Science Department, will illustrate:

Burning candles! Are we going back to all that brain-box stuff from the sixties now?...

It looks as if all these new ideas will take the fun out of teaching. If these materials are effectively going to teach science by some kind of osmosis, where does that leave me?

It is difficult to see how we can find time to fit all this Learning in Science Project stuff in. Our programme is too full already.

It is nevertheless possible to capitalise on negative reactions, which can lead to worthwhile discussion about:

• teaching methods which ignore children's ideas;
• the importance of the teacher's understanding of those ideas; and
• the doubtful value of teaching complex ideas based on faulty foundations.

To highlight some of the main propositions put forward in earlier chapters, and to provide some workshop discussion topics related to each of these, we set out below three general and four specific issues adapted from Tasker (1982b). It is assumed that those participating in the workshops would have previously been exposed to some of the findings about children's ideas discussed earlier and that they would therefore be aware of the discrepancies between teacher intentions and children's learning which often occur.

Problem 1

Teachers at all levels do not sufficiently appreciate the tenacity of existing ideas which learners hold and which permeate every aspect of science teaching.

> Children from a young age, and prior to learning science at school have meanings for words and views of the world which are to them sensible and useful. Such views can be strongly held and are often not recognised by teachers, but nevertheless influence the formal learning of science in many unintended ways.

Children have everyday meanings for words such as friction, animal, plant, living, rock and so on. All these words are commonly used in science but in that context have different meanings. Children also build up their own view of the world. The intuitive reasons for how and why things behave as they do may be quite contradictory to the scientific viewpoint, for example, force and motion, electricity flow in simple circuits, and the burning of a candle.

Questions for discussion
1   Why are most teachers unaware of children's ideas?
    What can we do to remedy this?
2   How significant is this problem in science and/or in other subjects areas of the curriculum?

Problem 2

Units of work, as well as individual lessons, are often planned without taking into account how pupils learn.

> It is not helpful to plan as though we, as teachers, are putting new ideas into otherwise blank minds. All learners have to construct meanings from what they see and hear, using existing ideas in long-term memory.

The way pupils think and what pupils learn is dominated by what they know already. It is useful to regard science teaching as an opportunity for individuals to change their children's science into scientists' science. Hence, for science teaching to be successful, teachers need to understand what is comprised in both children's science and scientists' science, and to recognise the various roles they can adopt to facilitate learning, for example, motivator, diagnostician, guide. In these ways pupils can be helped toward accepting new ideas as more intelligible, plausible, and useful than previously held ideas.

Questions for discussion
1   Is it true that teachers of science, and/or of other subjects, usually think
    in terms of placing pre-packaged meanings in pupils' minds, rather than
    in modifying the learner's existing ideas?
2   To what extent would a change of viewpoint about how children learn lead
    to changes in teacher's actions, roles, and strategies?

Problem 3
The planning of courses, units of work, and lessons, is not always based on
any clear overall objective.

> Teaching should help all pupils to make better sense of their world
> and to think about that world in more effective ways. Teaching should
> help pupils to see that some of the non-intuitive ideas which have been
> developed through the collective wisdom of our culture are indeed
> intelligible, plausible and useful to society. Teaching should help all
> pupils adopt some of these perspectives as their own.

Both teachers and pupils need to be continually reminded of the goals of
teaching and learning. Even if the aim is simply for pupils to pass an
examination, they are more likely to do so if what they are doing makes sense
to them, rather than blandly following instructions. But we must also accept
that pupils' priorities in terms of making sense of the world may not currently
be within our teaching domain, for example, the pop world, the world of sport,
the world of human relationships.

Questions for discussion
1   What should be the aim of teaching in science, and/or other subjects, in
    terms of changing children's ideas?
2   Are we always genuinely concerned to change the ideas of individual
    children in a way which is useful and productive to *them*?

Problem 4
Pupils often fail to see the connection between lessons in a sequence.

> Lessons are perceived by pupils as isolated events whereas to the
> teacher they are parts of a related series of experiences.

Lessons are usually part of a series which forms the classroom programme
for a particular topic. They are viewed by curriculum writers and teachers as
providing not only a particular experience for pupils and a specific influence
on their views, but also as contributing to a set of related experiences designed

to have a collective influence. Unfortunately, this broader perception of a single lesson is not recognised by many children.

This narrow focus of some pupils should be significant to teachers who draw heavily on previous classroom experiences when setting the scene for an activity. Links which a teacher perceives as strong and obvious, especially if they relate to the ideas taught in previous lessons rather than to what pupils did or what happened, may be far from obvious to pupils whose concern is with what they will have to do today. Another aspect of this problem is that teachers refer to the past in terms of their own perceptions, and unfortunately these perceptions are frequently different from those of their pupils.

Questions for discussion
1   What evidence is there to show whether or not this is a prevalent difficulty?
2   What methods can we use to overcome this problem with our own classes?

Problem 5
Pupils often fail to discern the objective of a specific lesson.

> The pupil-perceived purpose for the task is not the teacher-perceived purpose for the task.

Our research has shown that the teacher's perceptions of the purpose for a set task have often not been matched by those of his or her pupils. This has various consequences.

1   Some pupils who cannot immediately perceive a purpose for the lesson limit their involvement to following the instructions. They act in ways expected by the teacher and it is therefore an effective coping strategy, but has a very low level of intellectual involvement.
2   Some pupils establish their own alternative purposes, which the teacher in turn may, but usually does not, realise.
3   Some pupils who do not recognise the purpose, recognise instead that they have to provide the expected outcome and are thus concerned to 'get the right answer'.
4   Some teachers establish purposes for a lesson which are 'private' i.e. are not communicated to the student. These private purposes are frequently not achieved because pupils do not realise what is really required of them.

Questions for discussion
1   To what extent, again, is this problem a significant one?
2   How can we identify and then assist pupils who fail to discern our purposes?

Problem 6
Pupils frequently do not realise the importance of specific steps in an activity.

> Pupils' perceptions of the activity do not include an understanding of critical features in its design although teachers assume they recognise the importance of these.

To the teacher and curriculum writer the purpose of various features in science activities is usually clear and unambiguous. Such people have a 'feel' for the activity itself and the nature of the outcome. Pupils, on the other hand often have little feeling whatsoever for features in the design of an activity, and consequently little appreciation of what to expect in the way of outcomes.

Unfortunately teachers do not often sense this problem; they give instructions which contain critical features but almost never elaborate on their significance. This greatly increases the chance that pupils will be distracted from the significant outcome of the activity towards secondary outcomes.

Questions for discussion
1   Why should teachers spend more time explaining the reasons for what they want their pupils to do?
2   How can we use science activities to sharpen pupils' critical abilities in this area of human knowledge?

Problem 7
Pupils have difficulty relating what they do and see in science lessons to what they already 'know'.

> The significance of many activities is different for teachers and pupils.

Many factors operate to influence just what it is that pupils perceive as significant outcomes from a lesson. Some of these factors actually serve to reduce the influence of the classroom experience in altering the pupils' viewpoints.

Even in situations where pupils appreciate:

• how the lesson relates to previous lessons;
• the teacher's purpose for the activity;
• the design features of the activity;
• the background against which the activity is being conducted;

they still do not necessarily have their prior viewpoint influenced in the way the teacher intended, because of long-standing preconceptions about specific causes and effects.

Questions for discussion
1   What steps do we have to take before we can be reasonably sure that our teaching has had the effect we desired?
2   What techniques can you suggest for overcoming this problem in your classes?

Conclusions

There are no simple ways by which teachers can be sensitised to the issues we have raised in this book. What can be done, however, is to increase the number of teachers who appreciate the issues and who are actively attempting to find solutions to some aspects of the problem both within their own classrooms and through discussion with other teachers.

13 Epilogue

Peter Freyberg and Roger Osborne

Much has been learnt about children's science in the last few years. As references throughout this book show, there is now a substantial body of common knowledge, both of methodology and of topics investigated, gathered in many countries throughout the world. Considering the differences in the languages children use, the differences in everyday experiences which they meet, and the differences in the kinds of schooling which they receive, it is astonishing that the research findings are so similar. This encourages us to believe that children everywhere learn in similar ways and that, in many countries, they are exposed to similar everyday experiences and phenomena, both direct and vicarious, which influence their learning of classroom science. In addition, we are coming to realise how much children's responses to real-world science situations are guided by their own private ideas rather than by the propositions we have been at such pains to 'teach' them. Pupils may give us back the verbal formulae we have given them but when required to *act*, they often behave in accordance with their private ideas.

Research in children's ideas on science is continuing but it will take at least a decade, maybe two, to examine just those ideas children have which relate to even the main topics presently included in science syllabuses. What effects will these investigations have on what children actually learn? We cannot really say at this stage and it is likely that the additional insights gained over the next decade will require us to tackle much of the research all over again. Nevertheless, we believe that even the insights already obtained can dramatically influence current classroom practice. We do not think, however, that we should be seeking, or arguing for, one best teaching method or a definitive set of teaching objectives. Rather we need to seek ways to help teacher-colleagues and children themselves toward more effective learning in classrooms; perhaps taking into account some of the issues, and using some of the suggestions discussed in this book.

In helping pupils in their learning in science, and in encouraging other teachers to think about and take up some of the issues we have discussed, it may be profitable to consider our task in terms of some of the issues facing foreign-aid-project personnel working in cultures different from their own.

Effective aid personnel, like effective teachers, do not claim to be the inspired possessors of best techniques and methods. They have learnt that it is no use giving a ready-made irrigation system to nomadic tribespeople, nor a sanitation system to a group of forest-dwellers, simply because from a Western point of view these would provide obvious material benefits. Their approach is much more sensible and profitable. Firstly, they find out from the people themselves what problems most concern them. Then they help to isolate the real cause of the problem and explore with the people alternative ways of solving these problems. Only after this do they help the people construct for themselves what it is that they require. This may take more time than would be involved in a more directive approach, since changes in belief are usually involved as well as new information, but it avoids wasted effort and wasted resources. Further, the recipients are not left in a state of dependence on the aid given. We see this approach as having many similarities to the approach needed to change teachers' ideas about teaching and to change pupils' ideas in science.

Our experience with the Learning in Science Project, which was the starting point of this book, has been that many teachers need to be sensitised to what often happens in existing science lessons, and how children's ideas are being influenced, or not influenced, by schooling. Once teachers have convinced themselves that the existing conceptions of children are important — since such conceptions will inevitably be incorporated in future learning — their teaching frequently does alter, sometimes in small ways at first, to take greater account of children's present ideas. In this regard we see the main task of science teacher-educators, including senior teachers and curriculum-developers, as helping teachers:

- to confront the realities of classroom learning,
- to understand the importance of children's existing ideas,
- to understand children's ideas and,
- to realise how children's ideas compare with the views of scientists.

Most young children are, thank goodness, inherently curious. As we have seen they are only too willing to formulate theories about why the world is as it is. Our job as science teachers is to devise situations which sharpen up their ability to test those theories, which help them to assemble systematically the facts of the case before jumping to conclusions, and which highlight the consistencies and inconsistencies in their own explanations. Our job as science teacher-educators and curriculum-developers is to help teachers find increasingly more effective ways of tackling such problems.

Like all good aid projects, it is definitely a self-help and co-operative operation!

Appendix A

Finding Out What Children Think[1]

Beverley Bell, Roger Osborne and Ross Tasker

Throughout this book we have emphasised how important it is to find out what children think about a topic *before* we start teaching it. Sometimes there are research findings to guide us, often none. Even where research has something to tell us, it is always about other children, not our own. While many findings in science education research appear to be generalisable, across age-levels, across various curricula and across cultures (for example, children's ideas on causation), others are not. Thus, wherever possible, we should check our assumptions about children's prior knowledge with the situation in our own schools.

Interviewing children

Finding out what children *really* believe is not at all easy. Children spend a considerable portion of their childhood learning how to please their elders, and they are adept at fastening on small cues as to what is expected of them. In a teaching role, which even non-teachers adopt from time to time we are prone to use leading questions, to reject a 'wrong' answer by raising our eyebrows or rephrasing the question, and to praise the 'right' answer when we get it. Verbal and non-verbal cues such as these can quickly influence the authenticity of children's responses whether in a group or in a one-to-one situation. Effective probing of children's real ideas requires a conscious value-free approach which it is not at all easy for teachers to maintain — but maintain it we must if we are to get the information we need.

Whether we decide to proceed by interviewing individuals or by talking to small groups of children — and there are advantages and disadvantages in both these approaches — we have found it helpful to explain openly and clearly what we hope to achieve. For example:

Next week we are going to start a new series of lessons about various kinds of rocks. Before we do so, I want to make sure that we are all talking about the same thing. What do you mean when you talk about a 'rock'?...

---

1 This appendix has been adapted from Bell and Osborne (1981) and Tasker and Osborne (1981).

Children often talk so quickly that it is impossible to write down even a skeletal account of their responses so we invariably use an audio tape-recorder if at all possible. We turn the recorder on *after* the interview has begun, explaining that we wouldn't otherwise be able to remember everything that they said. We also tell them that we will remove anything from the tape that they are unhappy about, if they wish, thereby suggesting that we do not see it as a big issue. The recorder needs to be placed close to the child (children) as the adult's voice is almost always stronger than the children's and in any case we can fairly easily guess what we said even if it is not very distinct. We always try to sit *alongside*, rather than opposite, a child or group.

We have found the interviewing of individuals a surprisingly difficult business and we have learnt not to be too depressed especially with our first attempts. Even after more than 500 interviews as part of the original Learning in Science Project we were still not completely satisfied with some of our efforts. As Ravenette (1977) has suggested, the skill of the interviewer is to know when to ask a question and what question to ask. In the Interview-about-Instances (I.A.I.) approach, as with most interviews, the aim is to get the interviewee to talk. The interviewer chooses the topic, but what the interviewee says is most informative if it does not need to be extracted laboriously by a long series of interviewer questions followed by mono-syllabic answers. The aim of the interview is to get pupils to express their ideas in their own words. The interview is not an interrogation. The aim is to ask questions which show a genuine interest in the responses the interviewee makes and encourages children to respond further. Questions need to be easy to answer rather than difficult, neutral rather than leading, but on the other hand penetrating rather than superficial.

It is important to monitor one's own interviewing performances, as it is almost impossible otherwise to avoid the intrusion of such cues to children as we referred to earlier. Some problem areas, based on our own experience, and comments made by Simons (1981), are:

1   An interviewer needs continually to reiterate his or her stated interest in the child's meanings, and not go looking for an answer which will be assessed with respect to some external criterion. For example, consider:

Interviewer: *Do you know why the person might be able to see the candle?*

Interviewer: *Do you know how the eyes work?*

Interviewer: *Do you know what happens to the sunlight on the moon?*

These questions would have been much better if they had been phrased differently. For example:

Interviewer: Why *do you think* the person might be able to see the candle?

Interviewer: Can you explain to me the way *you think* the eyes work?

Interviewer: *What do you think* happens to the sunlight on the moon?

An interviewer's tone of voice, expression, emphasis and intonation are important considerations, as they need to be encouraging but not suggestive of expecting any particular answer. Often a nod, smile or 'a-hum' can be given to maintain communication and put the child at ease. However, care must be taken not to convey messages to the child other than those intended.

Teachers and ex-teachers have a particular problem when it comes to the neutrality within the I.A.I. technique. When teachers interact with students individually, they often almost unconsciously lead them through a series of questions which are aimed at developing a new student conception. This is the exact *opposite* of the I.A.I. procedure. *It is not the interviewer's conceptions that we are trying to get into the child's head but the child's conception that we are trying to get into the interviewer's head.*

This complete turn around requires the teacher to make a major change in orientation which some find very difficult to do initially, either lapsing back to a teaching mode or at least to a mode which is simply checking to see if the interviewee has the 'right' answer. Such a lapse is critically damaging to the tone of the interview. The interviewer is now seen by pupils as not really interested in their personal view. He or she is seen by the child as a teacher in disguise, giving an oral examination. On the other hand, if the interviewer can, both by word emphasis and mannerism, convey to the pupil that he or she is really interested in the person's view, whatever the view happens to be, then pupils respond and grow in confidence as the interview proceeds.

2   One of the advantages of the I.A.I. method is that the interview is a mixture of *closed questions*, for example 'In your meaning of the words electric current is there a current in the battery?' (which are simple to answer) and *open questions*, 'Why do you say that?' (which are penetrating). A balance of closed and open questions, of simple and difficult questions, of superficial and penetrating questions, of neutral and very specific questions, is most important. In this way it is possible to maintain pupil confidence but at the same time establish clearly the way the pupil thinks about the topic under discussion.

Follow-up questions are particularly crucial, but must not 'lead' the respondent. For example 'What happens to the light?' is a better question than 'Does the light stay around the candle?' The latter question has already made assumptions about the way the child views the situation. It also encourages a simple yes or no response. Such interview data is really of no value as evidence to support the proposition that a child has a particular point of view. He or she merely agreed or disagreed with the interviewer's view, which is quite a different matter.

3   The interview technique enables the reasons behind a student's initial answer to be explored by including supplementary and exploratory questions. Listen carefully to the answers given and follow them up until you are quite confident that you fully understand the response. For example:

Interviewer: *Does the mirror make light?*

Student   : *You can get your reflection. If you shine a torch in it, it'll make light.*

Interviewer: *What's a reflection?*

Student   : *Something you look at and it does whatever you do.*

4   A useful technique, particularly when one gets an unanticipated answer, is to repeat the answer back to the child, as if mulling it over. This has a dual purpose: (a) it checks that the child's response is audibly recorded on the tape, and (b) it allows both the child and the interviewer time to think about the answer.

An interviewer needs time to formulate a question to follow up a student response. Unanticipated responses are the high point of an I.A.I. interview and the interviewer can't afford not to follow them up. For example:

Interviewer: *Does the rainbow make light?*

Student   : *People say that God made it.*

Repeating the response, for example 'People say that God made it?, in a thoughtful tone of voice, also gives the child the opportunity to elaborate if she so wishes.

If this technique were used continually, then obviously the whole interview would become very stilted. On the other hand, when a response is repeated, it is *critical* that it is repeated exactly. Here is an example of poor interviewing:

Interviewer: *What is steam make of?*

Student   : *It is kinda like water?*

Interviewer: *It is water.*

Student   : *Yes.*

The pupil didn't say 'It is water' and there is *no* justification for the interviewer's definite statement in response. In I.A.I. we need children's own responses — not agreement, or disagreement, with an interviewer's comment.

5   This technique of repeating the response leads us to the more general issue of 'wait-time' — that is the time the interviewer waits for a response from the pupil. A successful interviewer has to be patient. Pupils need time to formulate a response. Do not butt in, if for no other reason than that the interviewer

must appear interested in *everything* the pupil has to say. On the other hand, it is possible to wait too long when no response is forthcoming. Experience helps one to judge how long a pupil requires to think a question through. Sometimes a little encouragement is required. For example:

Interviewer: *I just want to get your meaning. Remember, there are no right or wrong answers.*

Interviewer: *Let me try and put that question another way.*

Interviewer: *Well, let's leave that now. We might come back to it later.*

It is also important to realise that some children or indeed, most children from some cultures — will sometimes respond initially with a 'yes' simply to indicate that they understand the question. The interviewer must then wait for the *answer* to the question.

6 Some students express doubt and hesitation. This should then be explored by the interviewer. For example:

Student   : *I don't know.*

Interviewer: *You are not sure?*

Student   : *No.*

Interviewer: *Can you tell me what you are not sure about?*

Often when a student says 'I don't know' he or she has lapsed back into assuming that the interviewer is looking for the 'right' answer. Frequently just repeating the question by re-emphasising *'in your meaning of the word. . .'* is sufficient to overcome the problem.

7 Sometimes students will misinterpret or misunderstand a question. This is interesting in itself and the interviewer may wish to explore this. On the other hand the interview technique also enables the interviewer to clarify the question and clear up any misinterpretations.

Interviewer: *What happens to the light that it makes?*

Student   : *You use it to see with.*

Interviewer: *Does it move anywhere, or does it stay around the candle?*

Student   : *Stays around the candle.*

8 As stated earlier the most important response is often the unanticipated one for it indicates the child is thinking about things in quite a different way to the interviewer. Such responses need to be handled with delicacy as the aim is to appreciate the child's thoughts without distorting them with inappropriate questions.

If you do not understand a student's answer, do not ignore it but try to get at just what it is that the student is attempting to tell you. For example:

Interviewer: *I am not quite sure I understand what you are trying to tell me. Could you tell it to me another way?*

Inevitably, some unusual responses are not seen as such in the interview setting. The reading of the transcripts often highlights a point not obvious in the interview itself. For example:

Interviewer: *Could you describe how it is that the person can actually see the sun?*

Student   : *The same way as he could see the candle, except for, he couldn't go far enough away from the sun so he wouldn't see it. And when he's. . . unless the earth rotated quite fast he might.*

In this case the interviewer did not chase up on the comment about the earth's rotation. An appropriate question might have been:

Unless the earth rotated quite fast he might — can you explain what you mean by that?

Clarifying responses is awkward where a student gives a long reply, and one does not want to interrupt in case it disturbs the student's line of thought. The interviewer must try to remember the various probe questions which he or she will inevitably formulate during the long response, and introduce these probing questions into the interview when the student has stopped responding.

If lapses in questioning do occur, all the interviewer can do is to be sensitive to them and attempt to concentrate harder on the student's answers in the next interviews. Such concentration is extremely demanding, and for this reason not more than two interviews should be undertaken without a reasonable break. Good listening and questioning require hard and fast thinking!

9   Another reason why an interview is so demanding is that the interviewer needs to be sensitive to contradictory responses. These need to be explored fully, at every opportunity. For example:

Interviewer: *Is the grass living?*

Student   : *No, because it hasn't got a brain, doesn't eat.*

Interviewer: (Later) *Is a tree living?*

Student   : *Yes, it moves and feeds on particles in the air and needs water, it needs fertiliser.*

Interviewer: *You said the grass wasn't living and yet you say the tree is.*

Student   : *Oh, it* (the grass) *is just like a tree, needs water and it moves by growing.*

Interviewer: *So why did you say that it was not living* (before)*?*

Student   : *Because it wasn't like us.*

Naturally the interviewer has to keep in mind a student's earlier responses so that contradictions with respect to an earlier part of the interview can be picked up. This is another reason why not too many interviews should be attempted consecutively. It becomes increasingly difficult to remember if an earlier contradictory statement was made by the current interviewee or a previous one.

10 The interview technique also allows the opportunity for *students* to query the wording and meaning of a question. For example:

Interviewer: *Does a heater make light?*

Student   : *What kind of heater? One of those with orange bars?*

Interviewer: *Is the book living?*

Student   : (Pause) *I don't know what you mean.*

Interviewer: *Well, we'll start with another one. Is the boy living?*

Student   : *Yes.*

Be patient and supportive of this kind of questioning. It encourages students to see the interview as something different to the normal test situation.

11 The interview technique can be used with very young children. Since questioning is usually oral those with a limited vocabulary and perhaps reading difficulties should not experience any problems. In their replies children also give their responses orally and by gesture or facial expression. However, it should be pointed out that young children may interpret the pictures literally.

Interviewer: *Is the bird living?*

Student   : *Yes, living; if he was dead, he'd be lying on the ground and most probably be eaten by now.*

In the interview situation this kind of problem has to be expected from time to time. In this particular situation the dead/alive perspective was able to be replaced by careful questioning, directed towards the living/non-living distinction. Young students may also focus on unanticipated details in the diagrams.

Interviewer: *In your meaning of the word, is a cow an animal?*

Student   : *It has four legs and not two like a bird. It is an animal.*

Interviewer: *Is there anything else about the cow that tells you it's an animal?*

Student   : *Those things* (pupil points to the udder) *under there for feeding.*

Again the additional response may be interesting in itself.

12   Occasionally, despite all efforts to make the interview informal and non-threatening, a child will lose confidence rather than gain it as the interview proceeds. The child's responses tend to become mono-syllabic and the silences longer. It is best to discontinue the interview in such cases.

With the shy, withdrawn child there is undoubtedly a major problem with the I.A.I. technique. Our knowledge of learners' concepts and cognitive systems, using the technique, comes from what a child says or does. However, one who does not talk in an interview can not be categorised as knowing nothing. We have to accept this problem, and can but assume that the views of such children are not scientifically different from their more talkative peers. Subsequent survey techniques following interview work can check this to some extent.

13   It is essential to read the question on each interview card to the student, or in some other way to verbally identify the card you and the interviewee are discussing for a useful audio-tape record. When transcribing and analysing data an interviewer statement such as, 'Now, what do you think about this card/question?' is not helpful when you are not sure which card was actually being discussed at that time. This can be a particular trap for the unwary when an earlier card is reviewed to clarify an apparent contradiction in responses, because then the cards are being discussed out of their normal sequence. 'Let's compare your answers to these two cards', is inadequate. One needs to state something like 'Let's discuss these two cards — the one with the seagull on it and the one with the whale on it.' Again, where a child refers to a card but doesn't mention it by name the interviewer needs to make quite clear that it is *verbally* identified for the purposes of the audio-tape record. For example:

Student   : *If you go back to this card.*

Interviewer: *The one with the whale on it?*

Student   : *Yes.*

Interviewer: *Hm Hm.*

Student   : *Well, I think that is not an animal...*

14   While the one-to-one situation enables an interviewer to get some sort of response to every question asked, he or she must be sensitive to the possibility that a child may give just any answer, simply to avoid a silence. Subsequent questioning can investigate the depth of thinking upon which the answer is based. This is not a common problem, we believe, but the interviewer needs to be mindful of the possibility.

15   Sometimes a structured question or card will be intentionally passed over by the interviewer. This may be necessary if the pupil shows signs of exasperation at being asked *what obviously to him or her* is exactly the same question to which the reply is always the same. However, one can never assume that, if the card had been shown, the pupil would have definitely responded in the way predicted. Thus omission of a card or cards may make the comparative analysis of the data from different pupils very difficult.

To summarise
The interview-about-instances technique places a very heavy responsibility on the interviewer. She or he has to be skilled in the art of questioning and also knowledgeable in the content area under discussion to be able to assess pupils' responses immediately and make decisions about further questioning. Fortunately, most of us improve with practice. During a set of interviews on one particular topic, too, the interviewing becomes easier and more effective, since fewer answers are given which are completely unanticipated. Furthermore, in transcribing earlier sessions the interviewer learns from his or her mistakes.

For these reasons, if for no other, it is most desirable to transcribe tapes as soon as possible after an interview. This also has the advantage that the interviewer is more likely to remember what was said where the tape is indistinct. The interviewer should always listen to each tape, and transcribe it personally wherever possible, but be warned, it is a time-consuming task.

Finally, most novice interviewers find it helpful to get a more experienced interviewer to sit in with them during the third or fourth interview. That is, after they have first gained a little confidence! If the novice does one interview, and then the more experienced person does the next interview, this raises many useful discussion points. Alternatively, or in addition, it can be most helpful to get an experienced interviewer to read the novice's transcripts and point out leading questions, responses that should have been explored further and so on. No interview, however experienced the interviewer, is so good that it couldn't be improved, so there is no need to be embarrassed by your first mistakes!

We may have made it appear that effective interviewing is an impossible task, especially for the busy teacher. We hope you will not be discouraged, however; the important thing is to try for yourself to find out what children really think. After a few interviews, refer back to our check-list (Figure A1); it may be more meaningful, and hopefully helpful, at that stage.

Collecting information in classrooms

A great deal can be learnt about children's thinking — and attitudes — by observing their interaction with the teacher and amongst themselves during ordinary classroom activities. This kind of information-gathering should not be left to outside researchers: it should be facilitated for teachers within their own schools.

Figure A1   Checklist for Interviewers

Do's	Don'ts
1 Try to establish clearly how and what the *pupil* thinks. Emphasise it is the *pupil's* ideas that are important and are being explored.	Do not give any indication to the pupil of your meaning(s) for the word or appear to judge the pupil's response in terms of your meaning(s).
2 Provide a balance between open and closed questions and between simple and penetrating questions. In so doing, maintain and develop pupil confidence.	Do not ask leading questions. Do not ask the type of question where it is easy for the pupil to simply agree with whatever you say.
3 Listen carefully to the pupil's responses and follow up points which are not clear.	Do not rush on, e.g. to the next card, before thinking about the pupil's last response.
4 Where necessary to gain interviewer thinking time, or for the clarity of the audio-record, repeat the pupil response.	Do not respond with a modified version of the pupil response; repeat exactly what was said.
5 Give the pupil plenty of time to formulate a reply.	Do not rush but on the other hand do not exacerbate embarrassing silences.
6 Where pupils express doubt and hesitation encourage them to share their thinking.	Do not allow pupils to think that this is a test situation and there is a right answer required.
7 Be sensitive to possible misinterpretations of, or misunderstanding about, the initial question. Where appropriate explore this, and then clarify.	Do not make any assumptions about the way the pupil is thinking.
8 Be sensitive to the unanticipated response, and explore it carefully and with sensitivity.	Do not ignore responses you don't understand. Rather follow them up until you do understand.
9 Be sensitive to self-contradictory statements by the pupil.	Try not to forget earlier responses in the same interview.
10 Be supportive of a pupil querying the question you have asked, and in this and other ways, develop an informal atmosphere.	Do not let the interview become an interrogation rather than a friendly chat.

11 Read the question out loud to pupils.	Do not rely on pupils' reading ability.
12 Where all efforts to develop pupil confidence fail, abort the interview.	Do not proceed with an interview where the pupil becomes irrevocably withdrawn.
13 Verbally identify for the audio-record, the pupil's name, age and each card as it is introduced into the discussion.	Do not return to earlier cards without verbal identification for the audio-record.
14 Be sensitive to the possibility that pupils will give an answer simply to fill a silence.	Do not accept an answer without exploring the reasoning behind it.
15 Appreciate that a card omitted will result in missing data.	Make no assumption about the way a pupil would respond to a particular card.

A necessary pre-condition, of course, is that it is the students who are being observed — not the teacher. Teachers who expose their classes to this kind of observation must be assured that their performance *as teachers* is in no way being judged, although it is inevitable that what they say to their students, individually or as a class, will affect the thinking and responses of those students. Our experience has been that most teachers, while a bit hesitant until they are sure of the goodwill and non-judgemental approach of the observer, actually welcome the chance to find out what is really going on in children's minds as they participate in a lesson. Not once have we had an adverse reaction to the information so gathered — tactfully recorded, but no more so than one expects in other colleague interchanges.

Perhaps the easiest way to suggest how such observations should be handled is to describe how we proceeded in the Learning in Science Project. Our aim throughout was to focus on what the learners thought and did, with as little perturbation to the situation as possible.

1   As outside observers we always aimed to make ourselves as inconspicuous as possible, both in appearance and behaviour. We dressed informally, but not too much so, and we found it best *not* to be assumed by pupils to be an inspector (supervisor), visiting teacher, or even a teacher-in-training.

2   We started our observations with the class at the beginning of a lesson. The settling-in period which normally occurs at a change of lessons provided the ideal opportunity to enter the class and to locate ourselves within it without drawing much attention to ourselves.

3   As far as possible, we assumed the status of a pupil. This was done in a variety of ways. We preferred the teacher not to introduce us. If we felt the need to introduce ourselves to an individual student, we used our first names only. We tended to sit in the body of the class, usually asking students where there was a spare seat. We sought student permission to associate with them while they worked or while we talked to them about their work. In working with students we always referred to the teacher formally.

The following exemplifies what we would regard as a successful integration into a class at the beginning of a lesson. In this case the student directly in front of the researcher was curious about why the researcher was there.

Student   : *Hello.*

Researcher: *Hello.*

Student   : *Are you a checker?* (Inspector)

Researcher: *No.*

Student   : *Are you a student teacher?* (Teacher trainee)

Researcher: *No.*

Student   : *Why are you here?*

Researcher: *Oh, I just want to learn about science and Mr...*
                  (teacher) *said I could come today.*

Student   : *Oh... then you better come up and sit by me.*

4   We avoided public interaction with the teacher. To be seen talking informally and quietly to the teacher could have identified us as one of them as far as the students were concerned. When interaction was unavoidable it was done openly so that students could hear.

5   We placed any observation instruments, for example tape recorder or notebook, on the table from the beginning but handled them in a matter-of-fact way. We kept note-taking to a minimum. Our audio recorders were small, portable and encased. We found it best to use the recorder freely but to ensure that it was never the focus of attention. Once it was switched on we then ignored it as far as possible. When asked by students 'Why have you got a tape recorder?' we found the best response was to say that we were there to learn about science and did not want to miss anything. This was readily accepted, in our experience, perhaps because it complemented the kinds of questions we asked.

6   As far as possible we fitted our movements and actions to the stream of class behaviour and conformed to teacher directions. (That sometimes meant sitting down when the class was asked to do so.)

7   We were interested in student actions in an unobtrusive way. We tended not to question students too soon. Our experience was that, if we intervened with questions too early, it influenced subsequent behaviour. When talking with students we used the students' own idiom as far as possible.

8   When we asked questions we framed them in such a way as to elicit the views of the student. As we pointed out above, teachers almost unconsciously ask questions in a particular way to establish whether or not students hold specific views, and then to lead them to a new view. We found that it was all too easy for a researcher who has been a teacher to continue, or revert to, this style of questioning. In addition, if a student asked a question which would normally be directed to the teacher, we did not answer but rather implied that it was a question for the teacher to answer! We emphasised the view that we were there to find out about learning, about what students were doing and thinking. The following is an example of an effective probing of children's ideas.

Researcher: *What's happening when the crystal dissolves?*

Student 1 : *Dissolves and keeps going.*

Researcher: *Is the crystal still there?*

Student 2 : *Yep.*

Student 3 : *Yeah — just though.*

Researcher: *Just though?*

Student 4 : *Yeah...the colour's come out of it.*

Researcher: *The colour has come out of it?*

Student 4 : *Yeah...dissolved out of it.*

One cannot take for granted what children are actually doing, just from observing them. It is often necessary to probe actions as well as thinking.

Researcher: *How did you find all these* (features)*?*

Student   : *Some of them, I looked up the back at the index and the others I just looked at the countries.*

Researcher: *What did you do first? Look in the index or at...?*

Student   : *At the pages.*

Researcher: *Is that quicker or what?*

Student   : *Yeah, that's quicker — a lot quicker than using the index.*

Researcher: *What's hard about using the index?*

Student   : *I don't understand the North and the West.*

It is all too easy to slip into a teaching role and this soon becomes obvious, albeit subconsciously, to students. Such interviews yield little information, contrary to the main purpose of the exercise.

Researcher: *Which is the part you don't understand? This?*

Student   : *Yeah, these amounts are different in the test tubes.*

Researcher: *So what does the whole sentence say? From here?*

Student   : *The amounts in the test tube all weigh the same but the amounts are different in the test tubes.*

Researcher: *What do you mean by amounts?* (Silence)
(The researcher now begins the slide into teacher.)

Researcher: *How much ironsand did you need* (to weigh the specified amount)*?*

Student   : *A little bit.*

Researcher: *Where did it come up to?*
(The researcher is now firmly committed to teaching what amount means.)

Student   : *About there.*

Researcher: *All right, now how far up did the sawdust come?*

Student   : *Right up to the top.*

Researcher: *What about the water?*

Student   : *The middle.*

Researcher: *Right, so the amounts are different. . . the volumes are different.*

Finally, many of the comments about the structured interview discussed earlier apply equally well to the informal interview in the classroom setting.

Figure A2 provides a checklist of some do's and don'ts based on our experiences. The overriding aim in working in classrooms is, in our view, to be unobtrusive and to achieve acceptance by students at an almost adult-to-adult level. Naturally it is difficult for us to be certain just how students considered us, but our experience has been that, by adopting the procedures we have outlined, we were assigned a suitable role in the classroom — something like a potential adult student trying the subject out to see if he/she liked it! At least the off-task comments and actions made in our presence suggested that students saw no need to change their behaviour as a result of our presence.

Figure A2   Checklist for Classroom Observers

Do's	Don'ts
1 Dress informally.	Do not dress in a way which will suggest that you are an authority figure.
2 Start your observing at the beginning of a lesson.	Do not arrive late for a lesson.
3 Assume the status of a student but treat students on an adult-to-adult level.	Do not in any way act as a teacher or interact with students on a teacher-to-student level.
4 Interact solely with the students.	Do not interact with the teacher in the presence of students.
5 Use observation instruments quite openly, in a matter-of-fact way.	Do not turn the tape-recorder on and off or in any other way fiddle with the instrumentation during the lesson.
6 Fit in with the movements and actions of the class.	Do not be obtrusive, acting in a different manner to the students.
7 Interact with students at the minimum level necessary to obtain the required information.	Do not interact with students in a way which will alter the behaviour you wish to observe.
8 Use questions which elicit students' views, whatever those views might be.	Do not use teacher-type questions and hence assume a teaching role.

Appendix B

Constructing a Survey of 'Alternative' Views

Peter Freyberg and Roger Osborne

For reasons discussed in earlier chapters, we believe that the only way we can find out what children really think about a particular topic is to talk to them individually, using cards or other materials as prompts (the interview-about-instances or interview-about-events approach). This can be a time-consuming task, however, especially if we want to establish which of various children's ideas are most common. To do this we need to survey a sufficiently large number of children to ensure that our sample is reasonably representative of the age-group in our culture.

Pencil-and-paper surveys are an obvious way round this difficulty but they have their own drawbacks. Most often such 'tests' are constructed by teachers or researchers, with the 'correct' answer as the starting-point together with a number of 'distractors' designed — by the teacher or researcher — to include particular types of error. This may elicit responses to the *constructor's* ideas but it does not always bring out what are the *children's* ideas. To do that we need to employ in our survey ideas and phrases that children have expressed.

The range of ideas which children hold on most science topics is of course enormous, but (fortunately) we have found that on most topics the majority of children's ideas can be grouped into four or five categories. These categories will not necessarily exhaust the logical possibilities of response to a particular instance but they are the ones children actually use in their thinking about the world.

In constructing such surveys, the following procedures are useful:

1 Experiment with a variety of cards or stimulus events before selecting those which hold most interest for the children and bring forth the clearest responses.
2 Select the most commonly-employed categories of response from the initial interviews, using children's wording as exemplars wherever possible. Ask children individually from another group if they would want to suggest any other answer, and be prepared to add that if a new sufficiently-common response emerges.
3 Using this material, construct a word plus diagram question with multiple-choice responses based on the commonly-employed categories. Check with colleagues for clarity, avoidance of ambiguity, and possible overlap between responses.

4   Reinterview several individual pupils using the pencil-and-paper question. Probe the reasons for their choosing the response they did, particularly in relation to their earlier responses. Pay special attention to their interpretation of any diagram which is used. The meaning of arrows can easily be misunderstood (Schollum, 1983), as can questions involving sequence or simultaneity of events.

5   Revise or abandon the item, if necessary. Unless it is really eliciting the views of most children, it is worse than useless as it will be misleading.

Even where these procedures have been followed, we have still produced survey questions which were later shown to be inadequate. Moreover, children's responses to multiple-choice questions will inevitably differ sometimes from those they would give in an interview situation, which is more open-ended. Nevertheless, we have found that multiple-choice surveys, carefully constructed from children's responses, can give us a fairly reliable indication of the prevalence of various viewpoints, which can then be taken into account in devising teaching programmes (see Appendix D1). Longitudinal studies as well as cross-sectional studies of children at different age levels, also become more viable. As we have seen throughout this book, the trends thus revealed can tell us a great deal about the effectiveness of our teaching.

Appendix C

Checklists for the Science Teacher
Ross Tasker

Appendix C1: A checklist for planning science activities

1 General

Class level: 3 4 5 6 7   Ability: Low Mid High Mixed
No. in class: Work: individually_____ in pairs_____ groups of_____

---

2 Background Thinking

2.1 What in particular do I want to achieve? (Knowledge, manipulative skill, experimental design, attitude)

2.2 What will be the most effective kind of activity to achieve this? (Experimentation, demonstration, exploration, problem solving, project)

2.3 Exactly how will this activity achieve its purpose?

2.4 Are the pupils ready for this activity?
   (a) Do they have the background ideas that I expect them to have?

   (b) Do they have the required skills for the activity? (Manipulation, observation, classification, graph or data interpretation)

---

3 Setting up and Carrying out the Activity

3.1 How will I convey my purpose for the activity to the pupils?

3.2 What instructions will I need to give the pupils so that the activity will be carried out as I intend?

3.3 How will I present the instructions? (Blackboard, chart, list of tasks, diagram, flow charts, etc.)

3.4 What level of language will I need to use?

3.5 How can I highlight critical design features of the activity so that its main points will not be missed?

3.6 What opportunities need to be provided to learn or review prerequisite skills?

3.7   What equipment is needed?

3.8   How will the equipment be made available to the pupils?

3.9   How much time should the activity take?

4   Handling the Outcomes

4.1   What directions do I need to give pupils for recording what they are doing, what they observe and what they think it all means?

4.2   What will I do if there is a mismatch between my intended results and the pupils' results and thinking?

OR 4.2   How can other unexpected results and thinking be directed to those of the activity's purpose?

5   Using the Outcomes

5.1   Where will I go from here?

5.2   What alternative activities could I use to further challenge pupils' thinking if necessary?

Appendix C2: A checklist to evaluate a science activity

1   General
Evaluation by   SELF   COLLEAGUE

---

2   Background Thinking

2.1   What purpose for the activity did pupils adopt? (The teacher's or another?)

2.2   How was this shown? (Observation, questioning, individuals etc.)

2.3   Did the kind of activity suit the purpose?

2.4   Were the pupils ready for the activity? Did they have:
    (a)   appropriate background ideas?
    (b)   the skills required to carry out the activity as expected?

---

3   Carrying out the Activity

3.1   Did pupils have any trouble working out what to do? If so, what caused the trouble (language, comprehension, unfamiliar equipment, etc)?

3.2   Did pupils appear to really understand what they were doing? (Did pupils understand the design features of the activity?)

3.3   Were there any unexpected problems with using the equipment or materials?

3.4   Did pupils have the skills required to carry out the activity as expected?

3.5   Did the students have enough time to do the activity?

---

4   Handling the Outcomes

4.1   Did the pupils get their *own* results? What did they do with them?

4.2   Were the results 'expected' by the teacher?

4.3   How were the results and thinking of the pupils related to the purpose of the activity — did they draw their own conclusions?

4.4   Did pupil conclusions match those expected by the teacher? If not, was the teacher aware of the mismatch and were any such mismatches effectively considered by the class?

Appendix C3: A checklist for pupil evaluation of a science activity

Please answer the following questions in terms of how *you* see the activity.

1 What were you trying to find out or show?

2 Were the instructions easy to follow?

3 Which words did you find hard?

4 Did you know exactly what to do?

5 Was anything hard to do or understand?

6 If you asked for help, whom did you ask?

7 Why that person?

8 When you did each part of the activity, did you really know what you were doing?

9 If you talked to other pupils about the activity, which parts did you talk about?

10 What were the results *you* got from the activity? What did your results mean to you?

11 When you got a result, did you think it was what the teacher expected you to find out?

12 What did you do with your results?

13 What have the results to do with what you were trying to find out, show, or prove?

14 Did you really understand what you were doing during the activity?

Appendix D

Surveys and Interview Schedules

Appendix D1: A survey of some science-related ideas

The following questions are about the word 'animal'.

1 Is a cow an animal?
   (a) Yes
   (b) No

2 Is a person an animal?
   (a) Yes
   (b) No

3 Is a whale an animal?
   (a) Yes
   (b) No

4 Is a spider an animal?
   (a) Yes
   (b) No

5 Is a worm an animal?
   (a) Yes
   (b) No

The following questions are about the word 'living'.

6 Is a fire living?
   (a) Yes
   (b) No

7 Is a person living?
   (a) Yes
   (b) No

8 Is a moving car living?
   (a) Yes
   (b) No

The following questions are about the word 'plant'.

9 Is a carrot a plant?
   (a) Yes
   (b) No

10 Is a tree a plant?
   (a) Yes
   (b) No

Questions 11-15 are about electric current.

11 A torch has three batteries in it, as shown in the diagram.

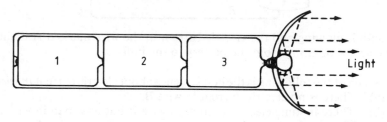

The torch is switched on and the lamp is glowing. Five students all have different ideas about the electric current through the batteries. Which one of the following ideas do you think is the best idea?
(a) No. 1 will have the most current.
(b) No. 2 will have the most current.
(c) No. 3 will have the most current.
(d) No. 1 and 3 will have more current than No. 2.
(e) They will all have the same current.

12 This question is about an ordinary electric light which is fixed to the ceiling. The light bulb has been taken out, but the switch on the wall is on.

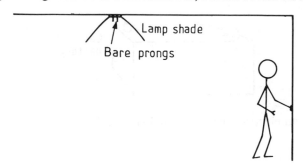

Is there an electric current in the bare prongs?
(a) No, because there can't be a current flowing.

    (b)   Yes, because if you touch it you get a shock.
    (c)   Yes, because if you put a bulb there it would glow.
    (d)   Yes, because the current would be going out from the prongs.

The following information is for questions 13 and 14.

A battery is connected to a torch bulb as shown.

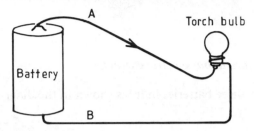

The bulb is glowing and the electric current in the wire marked A is shown by the arrow pointing from the battery to the bulb.

13   Which of the following is the best sentence about electric current *in wire B*?
    (a)   There is no electric current in wire B.
    (b)   There is some electric current in wire B but less than in wire A.
    (c)   There is the same electric current in wire B as in wire A.
    (d)   There is more electric current in wire B than in wire A.

14   Which of the following is the best sentence about the direction of electric current *in wire B*?
    (a)   The current has no direction as there is no current.
    (b)   The current is in the direction from the battery to the bulb.
    (c)   The current is in the direction from the bulb to the battery.

15   A car battery has been fully charged but has not yet been placed in the car. It is sitting on the bench in the garage and is not connected up to anything.

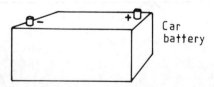

Is there an electric current in the battery?

16 Two metal rods are connected to the terminals on a battery. The rods are in a liquid as shown.

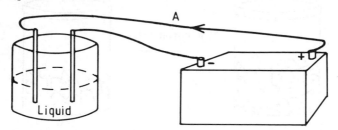

There is an electric current along wire A from the battery to the metal rod. Would there be an electric current in the *liquid*?
(a) It depends on what the liquid is.
(b) There must be a current in the liquid.
(c) There would not be a current in the liquid.

The following questions (17-19) are about things that happen in the kitchen.

17 When a kettle boils there are large bubbles in the water. What are the bubbles made of?
(a) Air
(b) Steam
(c) Heat
(d) Oxygen or hydrogen

18 If a wet saucer is left on the bench after it has been washed, then after a while it is all dry.

What happens to the water that doesn't drip onto the bench?
(a) It goes into the saucer.
(b) It just dries up and no longer exists as anything.
(c) It changes into oxygen and hydrogen in the air.
(d) It goes into the air as very small bits of water.

19 A small jar is filled with ice, the lid is screwed on tightly, and the outside of the glass is dried with a tea towel. Fifteen minutes later the outside of the jar is all wet.

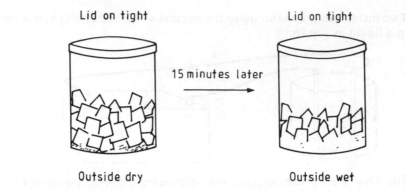

Lid on tight   Lid on tight

15 minutes later

Outside dry   Outside wet

Where has the water on the outside of the jar come from?
(a)   The water from the melted ice comes through the glass.
(b)   The coldness causes oxygen and hydrogen in the air to form water.
(c)   Water in the air sticks to the cold glass.
(d)   The coldness comes through the glass and turns to water.

20 The following diagram is a weather map. The big letter H mid-way between Australia and New Zealand shows:

(a)   high winds
(b)   high air pressure
(c)   hot temperatures
(d)   heat
(e)   humidity

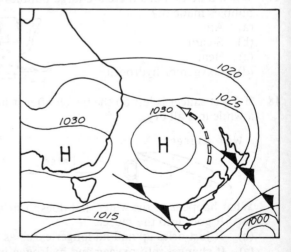

Appendix D2: Interview questions

1   Introduction

'I would like to talk to you about *your* meaning of the word animal. First, I'll show you some drawings and then we will have a chat about them.'

2   Key question for each card

'In your meaning of the word animal would you think of this as an animal?'

3   Follow up question for each card (choose one)

'Why did you say that?'

OR 'Can you explain to me why you think that?'

OR 'Can you tell me more about that?'

4   A final question

'Thank you for telling me about your meaning of the word animal. Just to finish I wonder if you could describe to me in your own words what · an animal is to you.'

A typical set of interview cards for animal could be the set shown in Figure 12.2.

Appendix D3: Survey investigating particle behaviour

The following questions are about the activity in which Condy's crystals (potassium permanganate) are dropped into two beakers, one of hot water and one of cold water.

1   Think of a Condy's crystal as it is before it is dropped into water. Your picture of the particles in Condy's crystals would be closest to which of the following drawings?

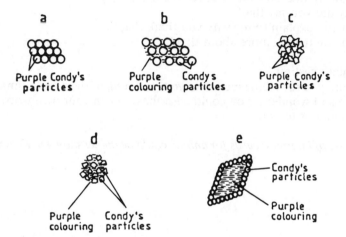

2   Your picture of pure water in its finest detail, would be most like which of the following drawings?

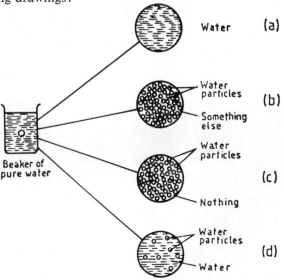

3   In your view, the purpose of this investigation was:
(a)   To find out the effect of heat on the behaviour of particles.
(b)   To find out if Condy's crystals will dissolve both in cold water and in hot water.
(c)   To find out if both hot and cold water are able to take away the colour from the particles that make up Condy's crystals and if hot water does it best.
(d)   To find out what happens when the particles making up a Condy's crystal move into solution.

4   How important are the following things in this experiment?
(Important, Not Important, Unsure)
(a)   The crystal needs to be coloured.
(b)   The two beakers used must be the same size.
(c)   The two beakers must be filled with the same amount of water.
(d)   One beaker must have hot water and the other beaker must have cold water.
(e)   The crystals added to the two beakers must be the same size.
(f)   The cold beaker must be kept well away from the hot beaker.
(g)   The two beakers have to be able to be seen through.
(h)   The crystals should be added to both beakers at the same time.
(i)   The crystals must not be added until the water in both beakers is settled.

5   Which of the following ideas best fits the way you think about what is happening when a Condy's crystal is dropped into the water?
(a)   The Condy's crystal sinks to the bottom and the purple colouring in it is slowly squeezed out by water pressure.
(b)   The particles of Condy's crystal go into the water as the crystal breaks up, and these Condy's particles make the water look purple.
(c)   Particles of Condy's crystal go into the water and allow the purple colouring to escape from inside the crystal into the water.

References

Andersson, B.   Some aspects of children's understanding of boiling point. In Archenhold, W. F., Driver, R. H., Orton, A., and Wood-Robinson, C. (Eds.)   *Cognitive Development Research in Science and Mathematics*. Leeds: University of Leeds, 1980.

Andersson, B. and Karrquist, C.   Electric circuits: Pupil perspectives. E.K.N.A. Project, Working Paper No.2, Department of Education, University of Gothenberg, Molndal, Sweden, 1979.

Andersson, B. and Karrquist, C.   How Swedish pupils, aged 12-15 years understand light and its properties. *European Journal of Science Education*, 1983, *5*, 4, 387-402.

Andersson, B. and Renstrom, L.   Oxidation of Steel Wool. Goteborg: E.K.N.A. — report No.6, Institute for praktisk pedagogik, Goteborg Universitet, 1982.

Ausubel, D.   *Educational Psychology*. N.Y.: Holt, Rinehart & Winston, 1968.

Angus, J. W.   Children's conceptions of the living world. *Australian Science Teachers Journal*, 1981, *27*, 3, 65-68.

Barnes, D.   *Language, the Learner and the School*. Hammondsworth, U.K.: Penguin, 1969.

Barnes, D.   *From Communication to Curriculum*. Hammondsworth, U.K.: Penguin Books, 1976.

Bell, B. F.   What is a plant; some children's ideas. *N.Z. Science Teacher*, 1981, *31*, 10-14 (a).

Bell, B. F.   When is an animal not an animal? *Journal of Biological Education*, 1981, *15*, 3, 213-218 (b).

Bell, B. F.   Teaching about animal, plant and living. Working paper No.31, Learning in Science Project. Hamilton, N.Z.; S.E.R.U., University of Waikato, 1981 (c).

Bell, B. and Barker, M.   Toward a scientific concept of animal. *Journal of Biological Education*, 1982, *16*, 3, 197-200.

Bell, B. and Osborne, R.   Interviewing Children. Working Paper No.45, Learning in Science Project. Hamilton, N.Z.; S.E.R.U., University of Waikato, 1981.

Biddulph, F.   Student's views on floating and sinking. Working Paper No.116, Learning in Science Project (Primary). Hamilton, N.Z.; S.E.R.U., University of Waikato, 1983.

Brook, A., Briggs, H. and Driver, R. Aspects of secondary students' understanding of the particulate nature of matter. Leeds: Children's Learning in Science Project, C.S.S.E., University of Leeds, 1984.

Brown, J. S. and VanLehn, K. Towards a generative theory of 'bugs'. In Carpenter, T. P., Moser, J. M. and Romberg, T. A. (Eds.) *Addition and Subtraction — a Cognitive Perspective*. New Jersey: Lawrence Erlbaum, 1982.

Brown, M. L. Place value in decimals. In Hart K. M. (Ed.) *Children's Understanding of Mathematics: 11-16*, London, John Murray, 1981.

Brumby, M. Problems in learning the concept of natural selection, *Journal of Biological Education*, 1979, *13*, 119-122.

Brumby, M. Students perceptions of the concept of life. *Science Education*, 1982, *66*, 4, 613-622.

Caramazza, A., McCloskey, M., and Green, B. Naive beliefs in sophisticated subjects: misconceptions about trajectories of objects. *Cognition*, 1981, *9*, 2, 117-123.

Carpenter, T. P. and Moser, J. M. The development of addition and subtraction problem solving skills. Paper presented to the Wingspread Conference, Racine, Wisconsin, 26-29 Nov, 1979.

Carr, K. C. Student beliefs about place value and decimals: Any relevance for science education? *Research in Science Education*, 1983, *13*, 105-109.

Carr, K. C. and Katterns, R. W. Does the number line help? *Mathematics in School*, 1984. (In press).

Carr, M. R. Concept names and mathematics learning. Hamilton, N.Z. Unpublished M.Ed. Thesis, University of Waikato, 1982.

Champagne, A. B., Gunstone, R. F. and Klopfer, L. E. Naive knowledge and science learning. *Research in Science and Technological Education*, 1983, *1*, 2, 173-183.

Champagne, A. B., Klopfer, L. E., and Anderson, J. H. Factors influencing the learning of classical mechanics. *American Journal of Physics*, 1980, *48*, 1074-1079.

Claxton, G. L. Teaching and acquiring scientific knowledge. In Keen, R. and Pope, M. (Eds.) *Kelly in the Classroom: Educational Applications of Personal Construct Psychology*. Montreal: Cybersystems, 1984.

Clement, J. Student preconceptions in introductory mechanics. *American Journal of Physics*, 1982, *50*, 1, 66-71.

Clement, J. Students' alternative conceptions in mechanics: a coherent system of preconceptions. In Helm, H. and Novak, J. D. (Eds.) *Proceedings of the International Seminar on Misconceptions in Science and Mathematics*. Ithaca, N.Y.: Department of Education, Cornell University, 1983.

Cockcroft, W. H.   *Mathematics Counts*. London: H.M.S.O., 1982.

Cohen, K., Eylon, B. and Ganiel, V.   Potential difference and current in simple electric circuits: a study of student concepts. *American Journal of Physics*, 1983, *51*, 5, 407-412.

Cosgrove, M.   *Mixtures — an Introduction to Chemistry*. Hamilton, N.Z. Hamilton Teachers College, 1982.

Cosgrove, M. and Osborne, R. J.   *Electrical Circuits — Teachers Guide*. Hamilton, N.Z.: S.E.R.U., University of Waikato, 1983.

Davis, R. B.   *Children's Learning of Mathematics*. Rome: Academic Nazionals Dei Lincei, 1979.

Deadman, J. A. and Kelly, P. J.   What do secondary school boys understand about evolution and heredity before they are taught the topics? *Journal of Biological Education*, 1978, *12*, 1, 7-15.

di Sessa, A. A.   Momentum flow as an alternative perspective in elementary mechanics. *American Journal of Physics*, 1980, *48*, 5, 365-369.

di Sessa, A. A.   Unlearning Aristotelian Physics: A study of knowledge based learning. *Cognitive Science*, 1982, *6*, 37-75.

Driver, R.   A response to a paper by Michael Shayer. In Archenhold, W. F., Driver, R. H., Orton, A., and Wood-Robinson, C. (Eds.) *Cognitive Development Research in Science and Mathematics*. Leeds: University of Leeds, 1980, 80-86.

Driver, R. H.   Pupils' alternative frameworks in science. *European Journal of Science Education*, 1981, *3*, 1, 93-101.

Driver, R.   Children's learning in science. *Educational Analysis*, 1982, *4*, 2, 69-79.

Driver, R.   *The pupil as scientist*. Milton Keynes: Open University, 1983.

Driver R. and Erickson, G.   Theories-in-Action: some theoretical and empirical issues in the study of students' conceptual frameworks in science. *Studies in Science Education*, 1983, *10*, 37-60.

Driver, R. and Russell, J.   An investigation in the idea of heat, temperature and change of state, of children between 8 and 14 years. Leeds: University of Leeds, 1982.

Duit, R.   Energy conceptions held by students and consequences for science teaching. In Helm, H. and Novak, J. D. (Eds.) *Proceedings of the International Seminar on Misconceptions in Science and Mathematics*. Ithaca, N.Y.: Dept. of Education, Cornell University, 1983.

Eaton, J. F., Anderson, C. W., and Smith, E. L.   *Student Preconceptions Interfere with Learning: Case Studies of Fifth Grade Students*. Michigan: Institute for Research on Teaching, Michigan State University, 1982.

Edwards, J. and Marland, P. Student thinking in a secondary biology classroom. *Research in Science Education*, 1982, *12*, 32-41.

Engel, E. and Driver, R. Investigating pupils' understanding of aspects of pressure. Leeds: Centre for Studies in Science Education, University of Leeds, 1981.

Erickson, G. L. Children's conceptions of heat and temperature. *Science Education*, 1979, *63*, 221-230.

Erickson, G. L. Children's viewpoints of heat: a second look. *Science Education*, 1980, *64*, 3, 323-336.

Erickson, G. L. Policy, perceptions and practice. New directions for science education. *Studies in Science Education*, 1982, *9*, 167-176.

Erlwanger, S. H. Case studies of children's conceptions of mathematics — Part I. *Journal of Children's Mathematical Behaviour*, 1975, *1*, 3, 157-283.

Fensham, P. J. A research base for new objectives of science teaching. *Research in Science Education*, 1980, *10*, 23-33.

Fensham, P. J. Misconceptions, preconceptions and alternative frameworks. (*Nyholm Lecture*). *Chemical Society Reviews*, 1984, *13*, 2, 199-217.

Flavell, J. H. On cognitive development. *Child Development*, 1982, *53*, 1-10.

Fowler, R. D. *The Language of Science*. London: Longmans, 1966.

Fredette, N. and Lochhead, J. Student conceptions of simple circuits. *The Physics Teacher*, 1980, *18*, 194-198.

Fredette, N. H. and Clement, J. L. Student misconceptions of an electric current: What do they mean? *Journal of College Science Teaching*, 1981, *10*, 280-285.

Freyberg, P. S. and Osborne, R. J. Who structures the curriculum: teacher or learner? *N.Z.C.E.R., SET*, 1981, 2, Item 6.

Gagne, R. M. and White, R. T. Memory structures and learning outcomes. *Review of Educational Research*, 1978, *48*, 187-272.

Gibran, K. *The Prophet*. London: Heinemann, 1926.

Gilbert, J. K., Osborne, R. J. and Fensham, P. J. Childrens' science and its consequences for teaching. *Science Education*, 1982, *66*, 4, 623-633.

Gilbert, J. and Pope, M. School children discussing energy. Mimeograph Report. Guildford, Surrey: I.E.D., University of Surrey, 1982.

Gilbert, J. K., Watts, D. M. and Osborne, R. J. Student conceptions of ideas in mechanics. *Physics Education*, 1982, *17*, 62-66.

Gilbert, J. K. and Watts, D. M.   Concepts, misconceptions and alternative conceptions: changing perspectives in science education. *Studies in Science Education*, 1983, *10*, 61-98.

Ginsburg, H.   Young children's informal knowledge of mathematics. *Journal of Children's Mathematical Behaviour*, 1975, *1*, 3, 63-156.

Glaser, B. G. and Strauss, A. L.   *The Discovery of Grounded Theory*. Chicago, Illinois: Aldine, 1967.

Goldberg, F. M. and McDermott, L. C.   Not all the many answers students give represent misconceptions: examples from interviews on geometrical optics. In Helm, H., and Novak, J. D. (Eds.) *Proceedings of the International Seminar on Misconceptions in Science and Mathematics*. Ithaca, N.Y.: Dept. of Education, Cornell University, 1983.

Guesne, E.   Lumiere et vision des objects. In Delacote, G. (Ed.) *Physics Teaching in Schools*. London: Taylor and Francis, 1978.

Gunstone, R. F., Champagne, A. B. and Klopfer, L. E.   Instruction for understanding: a case study. *Australian Science Teachers Journal*, 1981, *27*, 3, 27-32.

Gunstone, R. F. and White, R. T.   A matter of gravity. *Research in Science Education*, 1980, *10*, 35-44.

Gunstone, R. F. and White, R. T.   Understanding of gravity. *Science Education*, 1981, *65*, 291-299.

Happs, J. C.   Some aspects of student understanding of soil. *Australian Science Teachers Journal*, 1982, *28*, 3, 25-31 (a).

Happs, J. C.   Classifying rocks and minerals: a conceptual tug-of-war. *N.Z. Science Teacher*, 1982, *34*, 20-25 (b).

Happs, J. C.   Some aspects of student understanding of two New Zealand landforms. *N.Z. Science Teacher*, 1982, *32*, 4-12 (c).

Happs, J. C.   The utility of alternative knowledge frameworks in effecting conceptual change: some examples from the earth sciences. Hamilton, N.Z.: Unpublished D.Phil. Thesis, University of Waikato, 1984.

Hart, K. M. (Ed.)   *Children's Understanding of Mathematics: 11-16*. London: John Murray, 1981.

Hartel, H.   The electric circuit as a system: a new approach. *European Journal of Science Education*, 1982, *4*, 1, 45-55.

Hewson, P. W.   Aristotle: Alive and well in the classroom. *Australian Science Teachers Journal*, 1981, *27*, 3, 9-13 (a).

Hewson, P. W.   A conceptual change approach to learning science. *European Journal of Science Education*, 1981, *3*, 4, 383-396 (b).

Hodgson, B.   Girls in science: introduction. *Physics Education*, 1979, *14*, 5, 270.

Hurd, D. L. and Kipling, J. J. (Eds.)   *The Origin and Growth of Physical Science.* Hammondsworth, U. K.: Penguin, 1958.

Inagaki, K. and Hatano, G.   Collective scientific discovery by young children.   .
The Quarterly Newsletter of the Laboratory of Comparative Human Cognition, 1983, *5*, 1, 13-18.

Jones, A.   Investigation of students' understanding of space, velocity, and acceleration. *Research in Science Education*, 1983, *13*, 95-104.

Johnston, A. H.   Education: macro and microchemistry. *Chemistry in Britain*, 1982, *18*, 6, 409.

Kamii, C. and De Vries, R.   *Physical Knowledge in Pre-school Education.* Englewood Cliffs, N. J.: Prentice Hall, 1978.

Kargbo, D. B., Hobbs, E. D. and Erickson, G. L.   Children's beliefs about inherited characteristics. *Journal of Biological Education*, 1980, *14*, 137-146.

Karplus, R.   *Science Teaching and the Development of Reasoning.* Berkeley, California: University of California, Berkeley; 1977.

Karplus, R.   The learning cycle. In Collea F. P., Fuller R. G., Karplus R., Paldy L. G. and Renner J. W. *Workshop on Physics Teaching and the Development of Reasoning.* Stony Brook, N.Y.: American Association of Physics Teachers, 1975.

Katterns, R. W. and Carr, K. C.   Talking with seven year olds about times. *The Arithmetic Teacher*, 1984.

Kelly, G. A.   Ontological acceleration. In Maher, B. (Ed.) *Clinical Psychology and Personality: The selected papers of George Kelly.* N.Y.: Wiley, 1969.

Kingston, B. and Saville, S.   Reflections on a conference. *Physics Education*, 1983, *18*, 253-254.

Klein, C. A.   Children's concepts of the Earth and the Sun: a cross-cultural study. *Science Education*, 1982, *65*, 1, 95-107.

Knight, G. H.   A clinical study of the mathematical incompetence of some university students. Massey University, N.Z.: Unpublished Ph.D. thesis, 1983.

Lawler, R. W.   The progressive construction of mind. *Cognitive Science*, 1981, *5*, 1-30.

Layton, D.   *Science for the People.* London: George Allen and Unwin, 1973.

Leboutet-Barrell, L.   Concepts of mechanics among young people. *Physics Education*, 1976, *11*, 462-466.

McCloskey, M.   Intuitive Physics. *Scientific American*, 1983, *248*, 114-122.

McCloskey, M., Caramazza, A. and Green, B.   Curvilinear motion in the absence of external forces: Naive beliefs about the motion of objects. *Science*, 1980, *210*, 1139-1141.

Minstrell, J.   Explaining the 'at rest' condition of an object. *Physics Teacher*, 1982, *20*, 10-14.

Moser, J. M. and Carpenter, T. P.   Young children are good problem solvers. *The Arithmetic Teacher*, 1982, *13*, 3, 24-26.

Moyle, R.   Weather. Working Paper No.21, Learning in Science Project. Hamilton, N.Z.: S.E.R.U., University of Waikato, 1980.

Novick, S. and Nussbaum, J.   Junior high school pupils' understanding of the particulate nature of matter: an interview study. *Science Education*, 1978, *63*, 3, 273-282.

Novick, S. and Nussbaum, J.   Pupils' understanding of the particulate nature of matter: a cross-age study. *Science Education*, 1981, *65*, 2, 187-196.

Nussbaum, J.   Children's conception of the earth as a cosmic body: a cross-age study. *Science Education*, 1979, *63*, 1, 83-93.

Nussbaum, J. and Novak, J.   An assessment of children's concepts of the earth using structured interviews. *Science Education*, 1976, *60*, 535-550.

Nussbaum, J. and Novick, S.   The challenge of student alternative frameworks to teachers and researchers: a position paper. Jerusalem, Israel: Israel Science Teaching Centre, The Hebrew University of Jerusalem, 1981 (a).

Nussbaum, J. and Novick, S.   Brain storming in the classroom to invent a model: a case study. *School Science Review*, 1981, *62*, 221, 771-778 (b).

Nussbaum, J. and Novick, S.   A study of conceptual change in the classroom. Paper presented at the annual meeting of the National Association for Research in Science Teaching, Lake Geneva near Chicago, 1982.

Nussbaum, J. and Novick, S.   Alternative frameworks, conceptual conflict and accommodation: Toward a principled teaching stragety. *Instructional Science*, 1982, *11*, 183-200.

Ogborn, J.   Questions, questions, questions. In Ogborn, J. (Ed.) *Small group Teaching in Undergraduate Science*. London: Heinemann Educational, 1977.

Osborne, R.   Some aspects of students' views of the world. *Research in Science Education*, 1980, *10*, 11-18.

Osborne, R. J.   Children's ideas about electric current. *N.Z. Science Teacher*, 1981, *29*, 12-19.

Osborne, R. J.   Modifying childrens ideas about electric current. *Research in Science and Technological Education*, 1983, *1*, 1, 73-82.

Osborne, R., Bell, B. and Gilbert, J.   Science teaching and children's views of the world. *European Journal of Science Education*, 1983, *5*, 1, 1-14.

Osborne, R. J. and Cosgrove, M. M.   Children's conceptions of the changes of states of water. *Journal of Research in Science Teaching*, 1983, *20*, 9, 825-838.

Osborne, R. J., Cosgrove, M. M. and Schollum, B. W.   Chemistry and the Learning in Science project. *Chemistry in New Zealand*, 1982, *46*, 5, 104-107.

Osborne, R. and Gilbert, J.   A method for the investigation of concept understanding in science. *European Journal of Science Education*, 1980, *2*, 3, 311-321 (a).

Osborne, R. and Gilbert J.   A technique for exploring students' views of the world. *Physics Education*, 1980, *50*, 6, 376-379(b).

Osborne, R. J. and Schollum, B. W.   Coping in chemistry. *Australian Science Teachers Journal*, 1983, *29*, 1, 13-24.

Osborne, R., Schollum, B. and Russell, S.   Toward solutions: the work of the Chemistry Action-Research Group. Working Paper No.35, Learning in Science project. Hamilton, N.Z.: S.E.R.U., University of Waikato, 1982.

Osborne, R., Smythe, K., Biddulph, F. and Carr, K.   Research with a focus on the individual learner in the primary school. Symposium presented to the New Zealand Association for Research in Education Conference, Christchurch, December 1982.

Osborne, R. J., Tasker, C. R. and Schollum, B. W.   Video: Electric current. Working Paper No.52, Learning in Science Project. Hamilton, N.Z.: S.E.R.U., University of Waikato, 1982.

Osborne, R. J. and Wittrock, M. C.   Learning science: a generative process. *Science Education*, 1983, *67*, 4, 489-508.

Otterburn, M. K. and Nicholson, A. R.   The language of (CSE) mathematics. *Mathematics in Schools*, 1976, *5*, 5, 18-20.

Petchell, M. J.   *Science for the Seventies I, Mark Two*. Auckland, N.Z.: Coronet Books, 1976.

Piaget, J.   *The Child's Conception of the World*. London: Routledge & Kegan Paul, 1929.

Pope, M. and Gilbert, J.   Personal experience and the construction of knowledge in science. *Science Education*, 1983, *67*, 2, 173-203.

Posner, G. J., Strike, K. A., Hewson, P. W. and Gertzog, W. A. Accommodation of a scientific conception: toward a theory of conceptual change. *Science Education*, 1982, *66*, 2, 211-227.

Raven, R. J.   The development of the concept of momentum in Primary School Children. *Journal of Research in Science Teaching*, 1967-8, *5*, 216-233.

Ravenette, A. T.   Personal construct theory: an approach to the psychological investigation of children and young people. In Bannister, D. (Ed.) *New Perspectives in Personal Construct Theory*. London: Academic Press, 1977.

Renner, J.   The power of purpose. *Science Education*, 1982, *66*, 5, 709-716.

Rodrigues, D. M.   Notions of physical laws in childhood. *Science Education*, 1980, *64*, 1, 59-84.

Rowell, J. A. and Dawson, C. J.   Teaching about floating and sinking: an attempt to link cognitive psychology with classroom practice. *Science Education*, 1977, *61*, 2, 245-253.

Rowell, J. A. and Dawson, C. J.   Laboratory counter examples and the growth of understanding in science. *European Journal of Science Education*, 1983, *5*, 2, 203-215.

Russell, T. J.   Children's understanding of simple electric circuits. In Russell, T. J. and Sia, A. P. C. (Eds.) *Science and Mathematics, Concept Learning of South East Asian Children: Second Report on Phase II*, Glugar Malaysia: SEAMEO-RECSAM, 1980, 67-91.

Schollum, B. W.   Chemical Change. *N.Z. Science Teacher*, 1982, *33*, 5-9.

Schollum, B. W.   Arrows in science diagrams: help or hindrance for pupils. *Research in Science Education*, 1983, *13*, 45-59.

Schollum, B. W. and Happs, J. C.   Learner's views about burning. *Australian Science Teachers Journal*, 1982, *28*, 3, 84-88.

Schollum, B. W., Hill, G. and Osborne, R.   Teaching about force. Working Paper No.34, Learning in Science Project. Hamilton, N.Z.: S.E.R.U., University of Waikato, 1981.

Sere, M. G.   A study of some frameworks used by pupils aged 11-13 years in the interpretation of air pressure. *European Journal of Science Education*, 1982, *4*, 3, 299-309.

Shayer, M. and Adey, P.   *Toward a Science of Science Teaching*. London: Heinemann, 1981.

Shipstone, D. M.   A study of secondary school pupils understanding of current, voltage and resistance in simple D.C. circuits. Nottingham: Department of Education, University of Nottingham, 1982.

Simons, H.   Conversation piece: the practice of interviewing in case study research. In Adelman C. (Ed.) *Utterings and Mutterings*. London: Grant McIntyre, 1981.

Sjoberg, S. and Lie, S.   *Ideas about Force and Movement among Norwegian Pupils and Students*. Report 81-11, Institute of Physics Report Series, University of Oslo, 1981.

Smith, E. L. and Lott, G. W.   Teaching for conceptual change; some ways of going wrong. In Helm, H. and Novak, J. D. (Eds.) *Proceedings of the International Seminar on Misconceptions in Science and Mathematics*. Ithaca, N.Y.: Dept. of Education, Cornell University, 1983, 57-66.

Smythe, K. G.   The social studies concepts of children: customs and traditions. Research Series: Paper No.1, Hamilton, N.Z.: Waikato Social Studies Association, 1983.

Sneider, C. and Pulos, S.   Children's cosmographies: Understanding the Earth's shape and gravity. *Science Education*, 1983, *67*, 2, 205-221.

Snow, C. P.   *The Two Cultures and the Scientific Revolution*. Cambridge, U.K.: University Press, 1961.

Solomon, J.   *Teaching Children in the Laboratory*. London: Croom – Helm, 1980.

Solomon, J.   Learning about energy — how pupils think in two domains. *European Journal of Science Education*, 1983, *5*, 1, 49-59.

Stavy, R. and Berkovitz, B.   Cognitive conflict as a basis for teaching quantitative aspects of the concept of temperature. *Science Education*, 1980, *64*, 5, 679-692.

Stead (now Bell), B. F.   Living. Working Paper No.15. Learning in Science Project. Hamilton, N.Z.: S.E.R.U., University of Waikato, 1980.

Stead (now Bell), B. F.   Ecology, energy and the Form 1-4 science syllabus. *N.Z. Science Teacher*, 1981, *28*, 17-20.

Stead, B. F. and Osborne, R. J.   Exploring science students' concepts of light. *Australian Science Teachers Journal*, 1980, *26*, 3, 84-90.

Stead, K. E.   Outlook on Science. Working Paper No.50, Learning in Science Project. Hamilton. N.Z.: S.E.R.U., University of Waikato, 1982.

Stead, K. E. and Osborne, R. J.   What is friction? Some children's ideas. *Australian Science Teachers Journal*, 1981, *27*, 3, 51-57 (a).

Stead, K. E. and Osborne, R. J.   What is gravity: some children's ideas. *N.Z. Science Teacher*, 1981, *30*, 5-12 (b).

Strauss, S.   Cognitive development in school and out. *Cognition*, 1981, *30*, 295-300.

Tasker, C. R.   Some aspects of the students' view of doing science. *Research in Science Education*, 1980, *10*, 19-22.

Tasker, C. R.   Children's views and classroom experiences. *Australian Science Teachers Journal*, 1981, *27* 3 , 33-37.

Tasker, C. R.   Two lessons in one. *N.Z.C.E.R., SET*, 1982, *1*, Item 8 (a).

Tasker, R.   The learner's way and the teacher's way: a whole staff workshop on student learning difficulties. Working Paper No.55, Learning in Science Project. Hamilton, N.Z.: S.E.R.U., University of Waikato, 1982 (b).

Tasker, R. and Lambert, J. (Eds.)   Science Activities: specific problems; some solutions. Working Paper No.48, Learning in Science Project. Hamilton, N.Z.: S.E.R.U., University of Waikato, 1981.

Tasker, R. and Osborne, R.   Working in classrooms. Working Paper No.44, Learning in Science Project. Hamilton, N.Z.: S.E.R.U., University of Waikato, 1981.

Tasker, R. and Osborne, R.   Portraying pupils' classroom experiences. *Research in Science and Technological Education*, 1983, *1*, 2, 133-144.

Tamir, P., Gal-Choppin, R. and Nussemovitz, R.   How do intermediate and junior high school students conceptualise living and non-living? *Journal of Research in Science Teaching*, 1981, *18*, 3, 241-248.

Tiberghien, A.   Modes and conditions of learning. An example: The learning of some aspects of the concept of heat. In Archenhold, W. F., Driver, R. H., Orton, A. and Wood-Robin, C. (Eds.) *Cognitive Development Research in Science and Mathematics*. Leeds: University of Leeds, 1980.

Tiberghien, A. and Delacote, G.   Manipulations et representations de circuits electrique simples per des enfants de 7 & 12 ans. *Review Francais de Pedagogie*, 1976, *34*, 32-44.

Tinbergen, D. and Thorburn, P.   *Integrated Science. Bk II* (Wreake Valley Project). London: Edward Arnold, 1976.

Trowbridge, D. E. and McDermott, L. C.   Investigation of student understanding of the concept of velocity in one dimension. *American Journal of Physics*, 1980, *48*, 1020-1028.

Trowbridge, D. E. and McDermott, L. C.   Investigation of student understanding of the concept of acceleration. *American Journal of Physics*, 1981, *49*, 242-253.

Viennot, L.   Spontaneous reasoning in elementary dynamics. *European Journal of Science Education*, 1979, *1*, 2, 205-221.

Warren, J. W.   Energy and its carriers: a critical analysis. *Physics Education*, 1983, *18*, 209-212.

Watts, D. M.   Gravity — don't take it for granted! *Physics Education*, 1982, *17*, 116-121.

Watts, D. M.   A study of schoolchildren's alternative frameworks of the concept of force. *European Journal of Science Education*, 1983, *5*, 2, 217-230 (a).

Watts, D. M.   Some alternative views of energy. *Physics Education*, 1983, *18*, 213-217 (b).

Watts, D. M. and Gilbert, J. K.   Enigmas in school science: students' conceptions for scientifically associated words. *Research in Science and Technological Education*, 1983, *1*, 2, 161-171.

Watts, D. M. and Zylberstejn, A.   A survey of some children's ideas about force. *Physics Education*, 1981, *16*, 360-365.

West, L. H. T.   The researchers and their work. In Sutton, C. and West, L. (Eds.) *Investigating Children's Existing Ideas about Science*. Occasional Paper, School of Education, University of Leicester, April 1982.

West, L. H. T. and Pines, A. L.   How 'rational' is rationality? *Science Education*, 1983, *67*, 1, 37-39.

White, B. Y.   Sources of difficulty in understanding Newtonian dynamics. *Cognitive Science*, 1983.

Wittrock, M. C.   Learning as a generative process. *Educational Psychology*, 1974, *11*, 87-95.

Wittrock, M. C.   Learning as a generative process. In Wittrock, M. C. (Ed.) *Learning and Instruction*, Berkeley: McCutcheon, 1977.

Index